MAPPING THE LAND

Aerial Imagery
for Land Use
Information

James B. Campbell
Department of Geography
Virginia Polytechnic Institute
and State University
Blacksburg

RESOURCE PUBLICATIONS
IN GEOGRAPHY

Library of Congress Cataloging in Publication Data

Campbell, James B., 1944 —
  Mapping the land.
  (Resource publications in geography)
  Bibliography: p.
  1. Land use — Remote sensing. 2. Topographical
drawing. I. Title. II. Series.
HD108.8.C36 1983   621.36'78   83-9936
ISBN 0-89291-167-0

Publication Supported by the A.A.G.

Graphic Design by D. Sue Jones and CGK

Printed by Commercial Printing Inc.
State College, Pennsylvania

Cover: The land use and land cover pattern near Harrisburg, PA, 1:250,000. The cover was created from the digital USGS data base at the George F. Deasy Laboratory of GeoGraphics at Pennsylvania State University by Fred Irani and J. Ronald Eyton.

direct   15 Jn 84   Geol & Geog

Foreword and Dedication

Land resource use has long been an important theme in geographical research. Throughout the 20th century, geographers have applied contemporary field methods as well as newly developed remote sensing technologies to document and analyze patterns of land use. The skills and abilities of geographers in land use analysis have been applied in professional roles in universities, government, and industry. Challenges and opportunities in this area seem almost limitless, as we tackle increasingly complex issues of resource assessment, land use planning, and understanding linkages between land use and other social and environmental processes.

The use of remote sensing for land use mapping involves both theoretical and methodological issues, as well as development of personal skills. The former can be addressed in scholarly publications and technical manuals; the latter are not satisfactorily dealt with in words — experience becomes crucial. This volume, the fourth in the 1982 *Resource Publications in Geography* series, bridges the chasm between technical literature and the laboratory experience of the land use analyst. Here Jim Campbell provides a professional identity to the researcher, who is able to cast current work within longer-standing geographical traditions. He also offers a guide to the development of skills required for a continuing, competent role in land use analysis.

The advisory board of the *Resource Publications* would like this volume to memorialize James R. Anderson, who at the time of his death in 1980, served as Chief Geographer of the U.S. Geological Survey and as Vice-President and President-designate of the Association of American Geographers. During a career spanning three decades, Anderson served in academic posts at the Universities of Maryland and Florida and in government with the U.S. Department of Agriculture and the USGS. Anderson's work continues to provide guidance and leadership in the application of remote sensing to land use mapping. Anderson played a key role in the development of the USGS Land Use and Land Cover map series. The hierarchical system of land use classification created by Anderson and his staff has become an international standard. Anderson was also focal in bringing the long-term geographical tradition of land use mapping together with emerging remote sensing technologies.

This volume is dedicated to the memory of James R. Anderson, a scholar and leader among American geographers; to Anderson's colleagues and friends who continue his work; and to a new generation of geographers who will draw upon his contributions for years to come.

Resource Publications in Geography are monographs sponsored by the Association of American Geographers, a professional organization whose purpose is to advance studies in geography and to encourage the application of geographic research in education, government, and business. The series brings contemporary research in the various fields of geography to the attention of students and senior geographers, as well as to researchers in related fields. The ideas presented, of course, are the author's and do not imply AAG endorsement.

The editor and advisory board hope that this volume will remind us of the important traditions underlying contemporary research in remote sensing and land use mapping and will aid in the training of geographers in the fine tradition exemplified by Jim Anderson.

C. Gregory Knight, *The Pennsylvania State University*
Editor, Resource Publications in Geography

Preface and Acknowledgements

At Virginia Polytechnic Institute, where I teach courses in remote sensing, I have found that land cover mapping has formed a useful vehicle for teaching students how to interpret, organize, classify, and display information derived from aerial images. I have observed that students from many separate disciplines, working with all of the varied forms and formats of remotely sensed data, encounter essentially the same practical and conceptual problems in preparing maps and reports from aerial imagery. I also discovered that existing texts and teaching materials largely neglect those issues that present the greatest challenges to beginning students. In addition, there are few texts available that place current uses of aerial imagery for land use mapping in the context of the long tradition of geographic survey of land use patterns. Therefore, much of the information presented in this volume was developed to address these issues, in support of my classroom and laboratory activities in remote sensing courses.

In preparing this volume I have assumed that readers are students who are enrolled in, or who have completed, an introductory course in remote sensing. As a result, this volume does not discuss details of specific sensors, specific forms of imagery, or the interpretation of specific kinds of land use. These subjects have been covered in standard texts available to students. Instead, this volume focuses specifically on those general issues regarding the organization and presentation of land use information derived from aerial imagery. These concepts and principles apply equally to all forms of imagery, so their utility extends far beyond the examples used here as illustrations.

I have assumed also that readers are familiar with the land use classification developed by Anderson and colleagues (1976), or have easy access to this important document. Their volume is inexpensive, concise, readable, and easily available for classroom use, so there seemed to be no point in repeating in detail the information and ideas already presented so well.

Many of the ideas presented here were developed in teaching geography to university undergraduates, so I must acknowledge the contribution that their questions and suggestions have made to the conception and development of this volume. In addition, both my students and I have benefitted from conversations with Wayne Strickland (Chief of Land Use and Environmental Planning, Fifth District Planning Commission, Roanoke, VA), who has been generous with his time and knowledge, both in the classroom, and in his offices in Roanoke. With his help, many of the ideas have been tested and applied to areas near Roanoke. William Koelsch, Clark University, provided assistance in my search for correct bibliographic information. I appreciate the support of VPI's Spatial Data Analysis Laboratory (R.M. Haralick, Director) in some of the work that led to the writing of this volume. Glenn Best was helpful in assisting with many of the small but vital tasks required to complete this volume. John Crissinger, Geology Librarian at VPI, provided valuable support by orchestrating the interlibrary loan process to acquire, often on short notice, several valuable, if obscure, books and reports. C. Gregory Knight, as editor of the AAG Resource Publications in Geography, supported this volume through his encouragement, and his advice and guidance on editorial and production details. R.E. Witmer and his colleagues in the Geography Program, U.S. Geological Survey, were generous in their assistance with some issues relating to their work. J. Ronald Eyton provided useful criticism, as well as guidance on preparation of the cover.

Finally, I welcome the comments and suggestions of readers, including their ideas for further development of the ideas presented in this volume, their quibbles, and the errors they may discover.

James B. Campbell

Contents

List of Figures

List of Tables

1

The Domesday Book and Modern Land Use Surveys

The practical significance of land use information can be illustrated by an example ostensibly far removed from today's concerns, the events that followed the invasion of England by William, Duke of Normandy (William the Conqueror). In October of 1066 William's invading armies defeated the forces of Harold of Wessex in battle at Hastings. Harold had assumed the English crown in January, 1066, following the death of Edward the Confessor. Within months after William's victory he was crowned King of England, and in a few years, the Norman hold on England was firmly established. William brought great power and prestige to the throne of England and governed with an authority that extended to the most remote corners of the kingdom. Among the most immediate changes that accompanied Norman rule were expansions in the scope and efficiency of administration of governmental affairs (Brown 1968).

In 1085 William's agents began compilation of the Domesday Book, recording the results of a detailed survey of the lands and inhabitants of England. The name 'Domesday' means in this context 'doomsday' — day of judgment, referring to its role in assessment for taxation. This name, however, was apparently not used until many years after the survey was completed. Physically, the Domesday Book consists of two large volumes written in Latin; it exists today (in several versions) as England's oldest surviving public document. Scholarly interest in the Domesday Book began in the nineteenth century, and continues to the present. Considerable attention is directed to reconstruction of the details of the survey that produced the data (the Domesday Inquest), and to the administration, tabulation, and analysis of the results. Other scholars have attempted to use the information recorded in the survey to re-create the demographic and settlement patterns at the time of the survey (Darby 1952, 1977).

The Domesday Book

The Domesday Book was a record of land use and population patterns at a time long before most of us would have expected serious interest in such information. There are no surviving records that state the explicit purpose of the inquest, although it seems clear that it was compiled as a means of assessing resources subject to taxation.

The result is regarded as the most complete medieval survey in western Europe. Royal agents formed a kind of jury that accepted evidence from citizens and officials in each region. Citizens answered the most private questions concerning their land holdings. The book records a village-by-village survey of the entire country, with over

13,000 towns and jurisdictions listed. Apparently there were about seven circuits, or separate inquests, each covering at least three to five shires (counties). Analysis of the fragments of preliminary drafts that still exist suggests careful planning and execution of the inquest, including systematic editing, revision, indexing, checking, and supervision of workers by a professional staff. Errors and omissions are present, but the overall picture is one of efficiency and professionalism (Galbraith 1961).

The inquest recorded the names of manors, numbers of freemen, villeins, cotters, and slaves, resembling a modern demographic census. Also recorded were acreages of plowland, meadows, pastures, and the numbers and sizes of fishponds (Finn 1961). In this sense, the Domesday Book resembles a modern land use survey. The royal inquest also recorded ownership of land at three separate dates: (a) during the reign of Edward the Confessor (Harold's predescessor), (b) at the time of the Conquest, and (c) in 1086, at the time of the survey.

The Domesday Book, despite its apparent remoteness from modern concerns, is important in our consideration of the methods of collecting and the significance of using land use data. First, that the survey was conducted at all is remarkable. Medieval monarchs were not noted for their willingness to expend royal income on activities unlikely to contribute profit to their kingdoms. It must be concluded that William and his advisors regarded the data recorded in the Domesday Book as essential for efficient administration of England, likely to provide immediate practical benefits. Second, demographic and land use data were integrated into a single survey, so that the two forms of information are reported at comparable time periods for compatible aerial collection units. Many modern surveys aspire to integration of several forms of data in compatible formats, often with less success than the Domesday Inquest. Third, changes in land use were recorded, another feature of the Domesday Inquest desirable, but not always achieved, in modern land use surveys. Finally, the interest in accuracy is a precedent for modern concerns about the quality of data and information.

In the Domesday Inquest we can identify features worthy of consideration in our own efforts to map land use and land cover. Although the contexts and methods for modern surveys differ greatly from William's time, the Domesday Inquest can be regarded in some respects as a prototype for many of the qualities desired in today's surveys. Even if we could find no other attractive feature of the Domesday Inquest, we can aspire to match its speed of completion — to finish a survey of such scope within two years must be regarded as an achievement even worthy in modern times.

Modern Land Use Maps

William's data were tabulated in massive volumes, a rather unwieldy format for those interested in examining geographic patterns. Today we find maps to be a more concise and manageable vehicle for presenting similar data. Typically, a map of a region is subdivided into discrete parcels, each labeled as a single category. For many users, land use maps prepared in this manner appear to be the simplest of all maps:

(1) Informational content is straightforward; most people feel that they have a good grasp of the definition of 'agricultural,' or of 'urban' land, even if they have not specifically thought about land use before.

(2) Map content is subject to verification by direct observation.

(3) Unlike many other maps of similar appearance (such as the usual topographic or geologic maps), land use maps do not require specialized knowledge of the content, symbolization, and the cartographic model that forms the basis of map logic.

(4) Map format and symbolization are consistent with our most basic notions of what maps should be; the subdivision of a region into a mosaic of discrete, labeled parcels is one of the most easily understood mapping conventions.

(5) Likewise, the use of aerial imagery for land use mapping appears to be one of the most straightforward applications of remote sensing. Preparation of the land use map would seem to require only the delineation and labeling of categories easily recognized as they appear on images.

As a result, land use maps appear to be among the most basic to compile and interpret. Yet, simplicity of form and content conceal complexities that may emerge only as critical attention is devoted to the meaning and usefulness of specific maps. Once encountered, these complexities are found to be interwoven with the geographic patterns of land use, the classification system, and qualities of the aerial images used to make the map. Intricate errors in form and content may influence the usefulness of the map and are seldom subject to convenient correction, because they flow from decisions regarding map purpose, detail, scale, and use. The following paragraphs outline some concerns of researchers who prepare land use maps from aerial imagery. Most ideas discussed here will be more fully developed in subsequent chapters.

Generalization

The cartographic model used to portray land use patterns identifies each mapped parcel with a single category. In reality, most mapped parcels are composed of several categories of land use, the inevitable consequence of cartographic generalization. Generalization simplifies complex detail presented on aerial images into a form convenient for use by the map reader. Logical and visual generalization is, of course, inherent to all maps and is required for legible cartographic representation of land use patterns. Yet simplification also means that detail, perhaps important detail, is withheld from the map reader. For example, the astute reader of Figure 1 knows that parcel labels usually identify predominant land use within parcel boundaries, and that other, unspecified, categories ('impurities') are present. Yet no notice of the presence of these impurities is provided, detracting from the usefulness or the credibility of the map. The map reader lacks access to the original information from which the map was prepared, and must depend completely upon the map maker to present the information in a logical and consistent manner.

As a result, it is essential that the interpreter develop and apply a consistent strategy for generalization, tailored both to the aerial image and to the user's requirements, then clearly specify: (a) the existence of impurities in specific categories, (b) their approximate areal proportions, and (c) patterns of occurrence within the parcels.

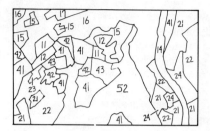 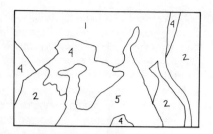

FIGURE 1   CARTOGRAPHIC GENERALIZATION. The detailed pattern (left) has been generalized (right) by combining categories and smoothing boundaries. Most land use maps exhibit some degree of generalization, so the labeled parcels include categories other than those identified by the parcel labels. As a result, land use maps must be accompanied by written material that describes the actual character of mapping units.

Compatibility With Other Data

By itself, a single land use map may have rather limited usefulness because the map reader often uses the map in relation to other data and information (Figure 2). Furthermore, the reader usually needs to accumulate land use data in a form that corresponds to the management and political units of interest and to other information required for the study. For example, land use data have maximum usefulness when they can be related to demographic and economic data for the same city or county. Comparisons can be meaningful only when land use data are compatible with other data in respect to date, category definitions, detail, spatial collection unit, and accuracy. If, for example, we wish to gather land use data in a form that can be related to census information, our land use data must be collected using imagery that corresponds at least approximately to the date of the census, shows detail comparable to that of the census data, and permits plotting of census sub-unit boundaries so that the land use data can be reported by the same geographic units as the census. Efficient preparation of a useful map depends, of course, upon early recognition of these requirements before the study is under way.

Assignment of Image Areas to Land Use Categories

Most aerial imagery presents a map-like representation of the landscape that seems to form a natural and convenient base for delineating land use. Usually, however, the photointerpreter is presented with much more information than can be accurately and legibly presented on a map. As a result, the interpreter defines a working model for relating detail on the image to specific land cover categories, then applies this model consistently throughout the image, using the model to assign areas of the image to categories in the classification system (Figure 3). This working model can be represented as a kind of filter that separates relevant from irrelevant detail or, alternatively, as a translator that can assign image detail to its correct informational category, much as a linguist assigns words and concepts in one language to corresponding words and concepts in another.

These filter/translators are devised and applied informally, almost intuitively, by photointerpreters in a manner that is tailored to the scale and resolution of the imagery, to the detail in the classification system, and to the publication scale of the final land use map. It is essential that each interpreter apply the strategy in a *disciplined, systematic,* and *consistent* manner throughout the image, so that the final map is uniform in accuracy and in representation of detail.

Map Characteristics Consistent with Ultimate Use

A land use map is usually prepared at the request of an institution or group of individuals. Although it may seem axiomatic that maps should be prepared for intended uses, maps often have inappropriate scales, levels of detail, categorization, or symbolization. There are an infinity of possible land use maps of an area, only a few of which may satisfy the user's requirements. Consider, for example, the varied maps that can result from alternative choices in respect to classification system, detail in the classification system, cartographic detail, date of imagery, and mapping scale. There is, of course, no single "correct" combination that is suitable for all circumstances. A poor choice in respect to even one of these qualities may render the final map unsuitable for its intended use. To add to these difficulties, it must be noted that the user's requirements may not always be self-apparent or obvious.

As a result, the photointerpreter/cartographer must question the map user at the outset of the project to outline alternatives and to assure that requirements for categorization, classification detail, spatial detail, and compatibility with other data are clearly understood and can be achieved with the resources at hand. As images are interpreted, and the map prepared, the interpreter should discuss progress with the user to detect problems at the earliest possible opportunity.

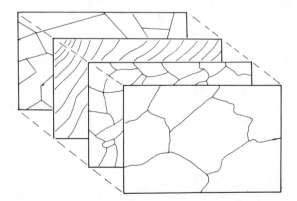

FIGURE 2   COMPATIBILITY OF VARIED FORMS OF DATA. Land use information has its greatest usefulness when it is compatible with other information (and with other land use data) in date, scale, accuracy, detail, and other qualities. Comparable information can then be superimposed (literally or figuratively) to form the kind of unified, multi-faceted data set suggested here. Several kinds of data can be integrated to provide a comprehensive view of a region. In reality, of course, such compatibility is seldom achieved without the explicit efforts of those who collect and manage information.

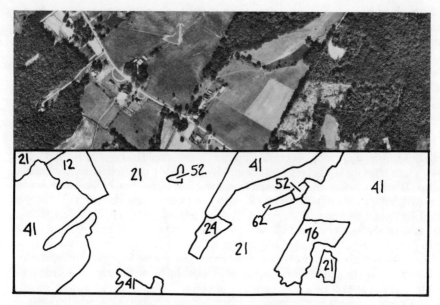

FIGURE 3   ASSIGNMENT OF IMAGE AREAS TO MAP CATEGORIES. The image interpreter must delineate areas of relative uniformity of land use, even though these areas may not always be uniform in appearance. This process requires accurate delineation of individual parcels, then correct assignment to land use categories. For complex landscapes, this process can be extremely difficult. Symbols here are keyed to Table 7.

Land Use Classification

Preparation of a land use map from aerial imagery is essentially a process of segmenting the image into a mosaic of parcels, with each parcel assigned to a land use class. A natural inclination for the novice interpreter is to devise land use classes based upon categories easily recognized and delineated from examination of the imagery. For example, 'suburban land' may seem to be a reasonable land use category because of identification ease and correspondence with informal land use classes developed from personal experience. In practice, of course, the most useful categories are those that match the informational needs of the map user. Typically categories such as "suburban land" are completely unsatisfactory for the user who requires division of land into functional classes such as 'residential,' 'commercial,' and 'industrial' land.

Because of the unpredictable interplay between image detail, classification detail, and map scale, the interpreter must find a balance between the precision of the classification system and the sizes of the parcels that can be interpreted, then portrayed legibly on a map. For example, a detailed category for cemeteries is of no practical value if all cemeteries in the mapped region occur in parcels of land too small to be legibly represented at the scale of the final map, or if the image has such coarse resolution that the cemeteries cannot be reliably interpreted. As a result, the image interpreter must prepare a classification system that is simultaneously compatible with the needs of the map user and consistent with image detail and map scale.

These problems, and others, render land use mapping a surprisingly difficult subject for many researchers, even those who have experience in related topics in remote sensing, cartography, planning, and other subjects. Many of the difficulties arise from conceptual, rather than purely technical, sources — concepts that are at the heart of the geographical approach to knowledge, related to appropriate choices of scale, generalization, and categorization for the purpose at hand. These topics are best studied in relation to a rather broad context that includes the uses, historical evolution, and guiding principles of land use inventory as applied by geographers and researchers in allied disciplines.

2

Land Use Information

William's agents probably went about their business without debating the meanings of definitions and concepts pertaining to land use survey. Their task was defined fairly clearly by others, and their attention was no doubt consumed by matters of immediate and practical significance. Today we must sort through an inventory of ideas, concepts, definitions, and meanings that may seem to have obscure distinctions, interesting only to the most meticulous scholar. Yet, because it is always important to have a good command of the ideas and concepts that form the foundation of our work, the following paragraphs outline some of the more important definitional issues pertaining to land use survey.

Concepts and Definitions

Land Use can be defined as the use of land by humans, usually with emphasis on the functional role of land in economic activities. Within this rather broad definition specific studies have employed a wide variety of subtle, often unstated, refinements in meaning (Burley 1961; Stamp 1966; Dickinson and Shaw 1977). There is little merit in a protracted enumeration of the varied connotations of the numerous terms that have been used, so the following discussion forms only a brief sketch of some basic issues.

Land use, as defined above, forms an abstraction, not always directly observable by even the closest inspection. We cannot see the actual use of a parcel of land, but only the physical artifacts of that use. Sometimes the implications of the artifacts are quite clear. A steel mill, for example, can be associated quite clearly with specific economic activities and land use categories. In contrast, a large extent of forested land may display little if any physical evidence of its varied uses, which might include production of timber, supply of water for distant urban areas, and space for recreation. In addition, some land areas may be characterized by contrasting activities (belonging perhaps to separate classes of land use) at separate seasons of the year. For example, some farmland might be used alternatively as cropland and as pasture at different times in the agricultural calendar.

The usual stereotype assigns a single, fixed, land use designation to each parcel. Often departures from them are not of great practical significance. We may be willing, for example, to accept designation of the predominant, or primary land use, or to define special categories for those multiple or sequential uses of significance. But it is important to have a firm grasp on the limitations of the conceptual models we use, so these issues deserve our attention even when they may not seem to have immediate practical significance.

Land Cover, in its narrowest sense, often designates the vegetative blanket, either natural or man-made, of the earth's surface. In a much broader sense, land cover designates the visible evidence of land use, to include both vegetative and non-vegetative features. In this meaning, dense forest, plowed land, urban structures, and paved parking lots all constitute land cover. Whereas land use is abstract, land cover is concrete and therefore is subject to direct observation.

Another distinction is that land cover lacks the emphasis upon economic function that is essential to the concept of land use. An hydrologist can focus interest solely to land cover because of a concern with the physical components of the landscape as they pertain to the movement of moisture. A traffic engineer who must use land use as an input to a traffic flow model must address the economic function of each parcel of land as a contributor of automobile traffic to the region's highways. Often the distinction between land use and land cover becomes more important as the scale of a study becomes larger and the level of detail becomes finer. Although many of these distinctions are significant, this volume, as a matter of convenience, will refer to 'land use' and 'land cover,' with full recognition that the terms do carry a wide range of meanings.

Land use is studied from many diverse viewpoints, by workers in pursuit of varied objectives. As a result, no single definition is appropriate in all contexts. Land use studies of interest in this volume are those that describe and inventory existing land uses of a region. This definition is necessary because much of the larger body of research pertaining to land use focuses upon the assessment of land capability and recommended land uses most appropriate given specific topographic, pedologic, climatic, and economic settings (for example, see Vink 1975). The kind of inventory of interest here examines existing land use, good or bad, without judgment. Such a report, if compiled objectively and professionally, forms the first phase in a more comprehensive process that can lead to subsequent analysis and interpretation. As a result, this volume addresses conceptual and practical problems encountered in reporting existing land use and land cover patterns.

More specifically, attention is devoted to the subject of compiling land use and land cover maps from aerial images (aerial photographs and other remotely sensed images). The maps that result can be considered as original information in the sense that they are generated by first hand examination of direct evidence of the land use patterns, rather than by compilation from secondary sources (such as census information or topographic maps).

The results are characterized by some degree of generalization because we are interested in mapping areas of relatively uniform land use and land cover, rather than individual objects. Typically, we do not wish to show the uses of specific buildings (except perhaps when they might occupy unusually large areas) but rather the major use to which a region of contiguous buildings is devoted. If we are consistent, and inform the map reader of our procedures, we may be willing to accept the inclusion of a few commercial buildings among those classified as residential. (Acceptable procedures for doing this are outlined in Chapter 5.) As a result, maps that form the subject of this volume must be considered members of the family of thematic maps, even though they may be presented at scales much larger than is usual for most thematic maps.

The Significance of Land Use Information

The Domesday Inquest forms only one example of the significance of land use information in governmental administration. A number of pivotal developments in the history of Western civilization have been interwoven with land use issues. For example, parliamentary enclosure of common lands in Britain (1750-1850; Hoskins 1955), the Proclamation Line in colonial North America, and the U.S. Homestead Act of 1864 are examples of major, far-reaching social and economic developments that have centered on land use issues. Almost all governmental units have a continuing requirement to create and implement laws and policies that directly or indirectly involve existing or future land use. In this context, it is surprising to realize that, until very recently, even the most modern governments were making these decisions from information that probably was inferior in both detail and scope to that collected by William's administrators.

Land use information is, of course, of great significance in scientific and scholarly research. National and regional land use patterns reflect the interaction between society and environment and the influence of distance and resource base upon mankind's basic economic activities. As a result, geographers, economists, and others have long regarded knowledge of regional land use patterns as fundamental in their studies of economic systems. Land use patterns are also recognized as influential elements in hydrological and meteorological processes. The importance of land use theory developed by von Thünen, Lösch, and others working in the disciplines of regional science, economics, and geography is evidence of the fundamental nature of land use in both theoretical and applied research. In a more immediate context, land use information is an essential consideration in the formation of public policies favoring orderly spatial development of economic activities at several scales (Clawson et al. 1960; Jackson 1981; de Neufville 1981).

Local Government Levels

If we exclude the possibility of annexation, each community has only a fixed amount of land area to allocate among varied economic and social activities required to support its citizens. There is increasing recognition that sensible use of finite, or possibly shrinking, resources requires comprehensive planning of community activities to coordinate the number and location of private and public facilities. Uncoordinated development can lead to unpleasant land use patterns, as well as to conditions that lead to undesirable environmental, social, and economic conditions. As existing development and increasing population limit options for use of land resources, development may be attracted to marginal areas or cause displacement of important activities from optimum sites. For example, residential or commercial development on steep slopes, on flood-prone areas, or on high-quality agricultural land may yield short-term benefits for some at the expense of long-term damage to the community as a whole.

In recognition of these problems, many states have legal requirements for local jurisdictions to prepare comprehensive plans outlining the kinds of land use patterns to be encouraged or discouraged on specific sites, as well as favored locations for specific uses. It must be noted that many such laws are now quite old (often they date from the 1930s or earlier), and strength of implementation varies widely. There can be no doubt that there is great variation in the quality of local plans and their administration. Yet in recent decades it has also become clear that some form of planning is essential to prevent uncontrolled distribution of economic activities from damaging environmental and human resources, and from disrupting efficient functioning of local economies.

The preparation of a comprehensive land use plan normally involves six steps (Table 1):

  (1) Data Collection and Research;
  (2) Analysis;
  (3) Plan Development;
  (4) Adoption;
  (5) Implementation; and
  (6) Review and Revision.

Land cover and land use data are of significance in the first step, which requires description and evaluation of existing land use. Historical surveys of land use changes over time can contribute to a knowledge of trends, and can permit anticipation of potential problems before they develop beyond the reach of local solutions.

Planners often collect land use and land cover information by direct observation from roads and highways — a procedure that may be satisfactory for some purposes, but may systematically exclude observation of inaccessible areas. Some jurisdictions have developed elaborate and detailed land use activity codes to permit systematic description of land use (Table 2). Accurate and precise application of these codes at fine levels of detail requires time-consuming effort by rather large staffs. For many areas, especially larger jurisdictions with rural and suburban land, use of remotely sensed imagery (especially aerial photography) may greatly improve the quality and detail of information available for the planning process.

The availability of suitable land cover maps can greatly alter the character of the decision making process. At the local level there are requirements for two kinds of data. The technical staff, in steps (2) and (3) of the planning process (Table 1) may require detailed information, including large scale maps with fine detail. The general public and the decision-making body may, on the other hand, require simplified maps and data tailored for visual and logical clarity. These conflicting requirements form a powerful justification for the use of hierarchical classifications (such as that proposed by Anderson *et al.* 1976). Information compiled at large scale, at fine levels of detail, can be easily generalized by collapsing detailed level III and II categories into fewer, broader categories for presentation at smaller scale and coarser detail.

Although attention may often be focused upon the technical use of land cover maps by the planning staff, these generalized maps may play an equally significant role in the planning process. During adoption, when a proposed plan is discussed and debated, the availability of suitable maps and data can greatly alter the character of the decision-making process. At public hearings the detailed maps so useful to the planning staff in developing the plan are usually of little value for presentation to a lay audience in a large meeting room or in printed matter designed to inform the general public. Simplified maps and data, presented with generalized patterns and classes, may alter the course of a meeting by providing a standard frame of reference for discussion and debate. Without such maps (often not routinely available at local levels of government), information regarding current land use patterns may be based solely upon personal observations of participants, which vary greatly in completeness and accuracy. Debates of controversial issues may be deflected from substantive issues to focus upon alternative perceptions of the same situation. The availability of maps provides a standard representation of the current situation and, therefore, can enhance

TABLE 1   THE COMPREHENSIVE PLANNING PROCESS

1. DATA COLLECTION AND RESEARCH
    Background Studies of Existing Conditions
    Trends and Forecasts
    Identification of Community Needs

2. ANALYSIS
    Evaluation of Data
    Develop and Assess Alternatives

3. PLAN DEVELOPMENT
    Prepare Community Goals, Objectives, Policies
    Integrate Components Into Comprehensive Plan

4. ADOPTION:
    Hearings, Decisions by Planning Commission, Governing Bodies

5. IMPLEMENTATION:
    Official Map
    Zoning Ordinance
    Subdivision Ordinance
    Capital Improvements Program
    Other Techniques

6. REVIEW AND REVISION

Source: Virginia Citizens Planning Association 1980.

the efficiency of the decision-making process. If the accuracy of the map is questioned, the debate can then focus upon the specific areas of dispute, which can presumably be resolved by further observation.

State and Regional Levels

Land use information for state and regional purposes differs greatly from uses by local governments. Planning processes at a state level often do not follow the kinds of established processes mentioned earlier for local governments, and seldom are their requirements equivalent to the state-imposed mandates followed by local jurisdictions. Nonetheless, land use information forms an important part of decisions made at the state level, especially in the formation of policy by executive agencies and of law by legislative bodies. For example, traffic flow models used to plan highway development at the state level require land use data as input to estimate traffic generated by neighborhoods supplying traffic to specific highways. State legislatures must often address issues regarding allocation of land to alternative uses, either in specific geographic regions (a decision to establish a state park or scenic reserve), or through general policies tailored for specific state-wide goals (laws to assist in preserving farmland from urbanization). In either context, the availability of accurate information regarding existing uses of the state's land is important in making sound decisions.

At the broad, statewide scale, the character of the required information differs greatly from that needed at local scales. For statewide land use and land cover information to be effective, it must be collected for the entire state at comparable levels of detail and accuracy, and at comparable dates. If compatible, hierarchial classifications have been used, it is theoretically possible to form a statewide data set from information collected at local levels (assuming the unlikely precondition that all component jurisdictions have collected information). In practice, however, it is unlikely that such an effort could be effective due to the innumerable practical problems in coordination and administration. As a result, discussion of statewide land use data must assume the availability of staff at the state level to assure compatibility and quality of information.

At local levels, it may be practical to gather data by direct observation, or to combine direct observation with the use of remote sensing imagery. At the state level it would seem to be impractical to rely upon any method except use of aerial photography or other remote sensing imagery. (Note, however, the exception to this recommendation in the Land Utilization Survey of Britain, described in the next chapter.) A number of states have prepared statewide land cover maps. The most effective means of conducting statewide inventories would appear to be establishment of some form of data base, since information is unlikely to be collected for the entire state at the same time. Revised and updated data would be an essential product for any statewide effort. These factors suggest that centralized coordination of any statewide mapping effort is essential.

The statewide inventories conducted in New York (Chapter 3) and in Minnesota represent examples of statewide land use information systems. Minnesota's Land Management Information System, administered by the University of Minnesota for the State Planning Agency, evolved from projects concerning land use decisions initiated in the 1960s (Borchert 1974; Hsu *et al.* 1975). It attempts to integrate land use information with demographic, resource, economic, and other information to provide accurate and timely information to individuals and institutions. Use of digital computers to store and analyze data provides opportunities to examine distributions at several scales and to examine interrelationships between separate distributions, capabilities usually impractical using other means of storing and representing data.

All such systems share practical and conceptual problems that prevent attainment of optimum capabilities, including variations in dates, accuracies, and levels of detail in source data, as well as use by individuals and institutions who do not appreciate limitations of the data. Nonetheless, it would seem that effective use of land cover information at the state level requires use of some form of data base, although possibly the most effective designs might differ greatly from those now in use. For example, it might be feasible and desirable to use systems with greater capabilities for interaction with local levels of government.

National Levels

At national levels, land use information is an important element in forming policies regarding economic, demographic, and environmental issues. In the United States, such policies might pertain to determining the location, extent, and character of surface mining, losses of agricultural land to urbanization, national parks and defense installations, or storage and disposal of hazardous wastes, to mention only a few of today's many controversial issues pertaining to land use. At this level, these issues are addressed (or often ignored) in executive actions of the various cabinet officers or in legislative action of Congress. Both types of policy must be formed using rather rudimentary forms of land use data. For example, examination of problems related to losses of prime agricultural land to urbanization is complicated by the varied statements of the severity and character of the problem. Two national programs for collecting land use and land cover data are described in Chapter 3. The Land Utilization Survey of Britain (1932-1938) formed a set of baseline data that have been widely used for both scholarly and administrative purposes. In the United States, the Geological Survey's Geography Program is compiling land cover and land use maps that will soon provide complete coverage of the 48 conterminous states. This information is organized in a form that permits reporting of land use by political and demographic units and by major drainage basins. These data should contribute to formation of national policy by providing a basis for defining problems from concrete information and by establishing criteria for making decisions.

International Levels

International requirements for land use data also focus upon many of today's major concerns considered at their broadest possible scales (NAS 1977). For example, major changes in land use within the world's major biomes (most notably the tropical forests, but also elsewhere) may have generated as yet unknown effects upon global biochemical cycles and upon the global energy balance. Other issues that require worldwide perspective include changes in global patterns of agricultural and forest lands, settlement patterns within zones of uncertain and variable climate, and efforts to control environmentally questionable agricultural practices.

Examination of these issues requires collection of data from many diverse sources — data that are compatible in scale, detail, accuracy, and categorization. For example, examination of land use within the rain forest, considered world-wide, requires data from many different regions, separated geographically, and with differing political and administrative traditions. Existing data, gathered independently by each nation, would probably be of minimal utility in a serious effort to examine issues of deforestation. Clearly an effort to coordinate details of separate surveys is required.

In 1949, a proposal to the International Geographical Congress (meeting then in Lisbon) outlined a strategy for mapping world land use patterns at a scale of 1:1,000,000 using standardized methods and co-ordinated approaches to classification and presentation. Although the Commission formed for this project agreed upon a standard classification and upon the broad outlines of the project, apparently only a few sheets of the survey were completed. Today the technical and administrative resources for such a project would seem to exceed those available in 1949, although the prospects for initiation and completion of a similar endeavor seem rather dim. Lounsbury and Aldrich (1979:6) acknowledged the substantial technical problems that the project faced, but observe also that it required ". . . a degree of international co-operation that has not existed in recent history."

Applications of Remote Sensing

Remotely sensed images lend themselves to accurate land cover and land use mapping, in part, because land cover information can be interpreted more or less directly from evidence visible on aerial images. Relatively little inference is required in most situations. It is possible to recognize two approaches, or traditions, in the analysis of land cover patterns as depicted on aerial photography.

There is evidence of early use of aerial photography for land use studies among geographers interested in relationships between environment and patterns of settlement and agriculture. Aerial photographs were recognized as one of the most dramatic and revealing means of examining patterns on the landscape. The American Geographical Society's *The Face of the Earth as Seen From the Air* (Lee 1922) represents recognition of the role of aerial photographs as concise representations of the complex patterns and interrelationships between cultural and environmental distributions. This volume represents an analytical tradition that examines land cover and settlement patterns, as depicted upon aerial images, as artifacts of the interplay between physical environment and cultrual history within a region. Such analyses are often subjective, qualitative studies of areas representative of larger regions. The essay by Gutkind (1956) forms another excellent example of this tradition.

A separate, more recent tradition can be described loosely as the use of aerial photography to conduct an inventory of the land use of a specified region, often as an element of a planning or development process. The use of aerial photography for this purpose apparently developed rather late, relative to many other applications of aerial photography such as topographic mapping or soil survey, for example. Although aerial photography was used for land cover mapping early after World War II, systematic assessment of the procedures and methods did not appear in the professional literature until the later 1950s and early 1960s (Anderson 1961).

Use of aerial photography as a means of regional survey does not necessarily conflict with the first tradition, but does tend to depend to a greater extent upon application of specialized technique and method — especially upon consistency of technique. The result is a set of maps and data tailored to meet the specific requirements of a user. It is this second tradition that forms the subject of this volume.

Land use maps are routinely prepared at a wide variety of scales, typically ranging from 1:12,500 to 1:250,000. At one end of this spectrum (the large scale maps), remotely sensed imagery may itself contribute relatively little information to the survey. Its main role may be to form a highly detailed base for recording data gathered by other means. At such large scales the land use map may actually form a kind of reference map, having little cartographic generalization. Such products are often used at the lowest levels of local government, mainly in urban areas, that both require such detailed information and have financial resources to acquire it.

As the scale of the survey becomes broader (as map scale becomes smaller), the contribution of the image to the informational content of the map becomes greater, although even at the smallest scales there must always be some contribution from collateral (non-image) information. Differing scales and levels of detail serve different purposes and different users. For the regional planner, the loss of resolution and detail at smaller scales may actually be an advantage when there occurs an integration and simplification of information that must be examined. For the medium- and small-scale land use surveys used in local planning, the product is a thematic map that depicts the predominant land cover within relatively homogeneous areas, whose delineation is subject to limitations of scale, resolution, generalization, and other constraints.

In recent years, geographers and others have identified and solved many of the more immediate practical problems encountered in systematic land cover surveys based upon remote sensing imagery. Both practical and conceptual issues in land use and land cover classification have been discussed in some detail. The U.S. Geological Survey's classification system (Anderson *et al.* 1976) seems to have gained widespread acceptance as a basis for small scale land cover mapping and as a framework for developing classifications at larger scales. Computer-based data systems tailored for the storage, manipulation, revision, and display of large amounts of land use data have been developed and applied in operational contexts. Procedures for evaluation and testing of accuracies of land use and land cover maps have yet to be perfected, although there has been considerable research on this topic.

Advantages and Disadvantages of Aerial Imagery

Use of remote sensing is not equally effective in all circumstances, so it is useful to enumerate the primary advantages and disadvantages of employing remotely sensed data for compiling land use maps. Advantages relative to most ground-based surveys are:

(1) Use of remotely sensed data may be cost-effective. Purchase of imagery and other materials may require only small or modest expense. Good results can be achieved by staff with only modest experience, if personnel are diligent and properly supervised.

(2) Speed may be an advantage if the area to be mapped is large, or if access to some areas is difficult. Once data and imagery are in hand, maps can be prepared quickly; revisions, updates, and corrections can be completed efficiently.

(3) In some areas, accuracy may be very good and can be checked by re-examining the aerial images. Time consuming field work can be focused upon problem areas that may be identifiable beforehand from preliminary examination of images.

(4) Aerial images form an historical record of land cover patterns at previous dates, so maps of land cover changes can be prepared from archival images.

(5) Maps generated by analysis of remotely sensed data typically exhibit some degree of generalization. The interpreter is required to identify and delineate areas of relatively uniform land use. The degree of generalization must be tailored to user requirements, but it is clear that generalization is usually desirable, generalization that is very difficult to achieve using ground observations.

(6) Aerial images inherently have a map-like format that lends itself to compilation of maps and to other operations less easily undertaken with other forms of data.

Disadvantages and limitations to the use of remotely sensed data can also be enumerated:

(1) Conventional ground-based land use survey is often conducted at levels of detail too fine to be compatible with detail interpreted from usual remote sensing images. Land use coding can extend to detail corresponding to individual lots and buildings (Table 2). If such detail is required, use of remotely sensed data may not be appropriate, or may be best used as a supplement to other methods. (Aerial images can, of course, be acquired at very high levels of detail which would not usually exploit the advantages of the imagery.)

TABLE 2   EXAMPLES OF LAND USE ACTIVITY CODES

TWO-DIGIT ACTIVITY CODES
  0 Housing:
    01 Household Units
    02 Transient lodgings
    04 Group quarters
    09 Other Residential
  1 Wholesaling, Warehousing, and Storage:
    10 Wholesalers with Stocks
    11 Warehousing and Storage
  2 Manufacturing (Non-Durable):
    21 Food and Kindred Products
    22 Textile Mill Products
THREE-, FOUR-DIGIT ACTIVITY CODES
    68 Wholesalers Without Stock:
      683 Dry goods and apparel
        6832 Dry goods, piece goods
        6835 Apparel and accessories
        6835 Footware
        6829 Others

Source: Scott *et al.* 1972.

(2) Financial constraints may require use of archived imagery or images acquired for other purposes. Image quality may not suit the immediate use. If it is necessary to map large areas, it may be especially difficult to acquire archival imagery for the entire region with compatible dates, scales, and qualities.

(3) Organizations may experience delays in identifying and acquiring appropriate imagery. The search for imagery may require contact with numerous organizations, and the process of identifying potentially useful imagery, then assessing coverage and quality, may require significant amounts of time. Delays while images are reproduced may be significant.

(4) Although remote sensing imagery can often be used with a minimum of equipment, even the most modest requirements of trained staff and space may prevent use in some situations.

(5) Land cover is, of course, often used as a surrogate for land use when aerial images are used. Although the distinction between the two may often be irrelevant for many practical purposes, it assumes greater significance as mapping scales become larger and classifications become more detailed.

3

Land Use Inventory in Retrospect

Today's land use and land cover maps have a characteristic form and organization that seems to be an obvious, simple, and straightforward means of mapping landscapes. The maps seem so familiar and simple that we tend to accept this form as the only, or the 'natural,' or the 'right' way to make such maps. In fact, map characteristics are not fixed, and are not the only way to organize land use information in map form. The maps that we use today have evolved from previous forms. In this chapter, we outline the evolution of land use and land cover maps and illustrate changes in respect to map purpose, form, and organization. We focus upon the evolution of the kinds of broad-scale land use/land cover surveys that have been of greatest interest to geographers and those in allied fields. One purpose is to place the characteristics of current maps and data into historical context and to relate today's maps to their predecessors. Our present maps, often prepared with the use of remotely sensed imagery, continue some features of early maps and data, which were, of course, compiled from direct observation from the ground.

It is difficult to define a clear beginning for the practice of land use mapping. Wallis (1981) identified an origin in the property surveys of large estates in England and France during the sixteenth and seventeenth centuries. She also defined links to cadastral and topographic mapping, and to city plans and insurance maps of the nineteenth century. Thomas Milne's map of London and nearby areas (1800) is often cited as the first 'true' land use map. His map, at two inches to the mile, anticipated many features of modern land use maps, including some current conventions in the uses of colors and symbols to depict land use regions.

Other origins for current land use maps can be identified in many of the thematic maps that came into use in the early nineteenth century (Robinson 1982). In thematic maps showing both demographic and physical distributions, we can see precedents for the logical organization and the symbolization used in modern land use maps. The broad-scale land use studies of interest to many geographers have, perhaps, an origin in the regional studies represented so well by the work of O.E. Baker. His map, *Agricultural Regions of North America* (Baker 1926), is not a land use map in the usual meaning of the term, but does represent the kind of regional analysis from which many aspects of modern land use maps have evolved. This map, one of a series by several authors on world agricultural patterns, is a kind of specialized land use map — a map of agricultural land use, compiled by integrating cultural, economic, physical, and census information. Baker showed 18 categories, mapped in rather broad detail on a small-scale map of North America. The map, despite its differences from modern land use maps, can be seen as a kind of model for our present maps in respect to overall purpose and organization. Although Baker used only 18 categories at a single level of detail, he recognized in his text that each of these is, in fact, the most general category of a hierarchy within each region. Baker's map can be seen as an example of a kind of geographic approach that is an ancestor of today's more specialized land use maps.

The Land Utilization Survey of Great Britain

The Land Utilization Survey was conducted during 1931-1938 under the direction of noted geographer L. Dudley Stamp. According to Stamp's accounts, there had long been interest among leading British geographers for some sort of national land use survey of England, Wales, and Scotland (Stamp 1960; 1962). Apparently there was support for the idea among academic geographers and from the geographical societies during the early years of this century. Stamp also noted concern for changes in the British landscape resulting from the depressed economic conditions in the late 1920s, especially in rural areas. Agricultural land abandonment and accompanying landscape changes were especially disturbing to those accustomed to the previous rural setting.

The Land Utilization Survey was funded in part by research funds from the London School of Economics (indirectly supported by a grant from the Rockefeller Foundation). Very little financial support was available, and much of the work was completed through volunteer effort. Support also was provided by the Ordnance Survey, the Geographical Association, and the Education Association. Other governmental agencies provided non-financial support and advice.

Stamp (1951:374) described the survey as ". . . a national stocktaking of land resources, using the methods familiar to all geographers, relying essentially on field work and direct observation." The entire land area of England, Wales, and Scotland was mapped using volunteer observers who recorded the land use using Ordnance Survey topographic maps as a base. Volunteers were recruited through educational and governmental agencies or through a descriptive brochure outlining the project. At the time, the entire country was covered by a map series at 1:10,560 (six inches to the mile). Each sheet represented six square miles, and depicted field boundaries, buildings, and other cultural detail. Approximately 22,000 sheets were required to represent the land area of England, Wales, and Scotland, so about 22,000 volunteers were necessary. Observers recorded the land cover of each parcel using the classification system devised for the survey (Table 3).

Coleman (1980) noted the unique coincidence of conditions favoring application of this approach in Britain. Britain has complete coverage by large scale topographic maps which depict field boundaries and other cultural detail. The vast majority of the land area is easily accessible to direct observation from the ground without long or arduous travel. The population provides a large pool of capable and willing volunteers. Even today there are few large areas that possess these characteristics.

The completed survey depicts conditions observed during 1931-1932. Some remote areas were not surveyed until 1938, but the majority of the field work was completed by 1932. Once field work for a sheet was done, manuscript maps were checked (sometimes in the field) for accuracy and for agreement at the edges. Information was reduced for presentation at 1:63,360. At this scale, only 140 sheets covered most of Britain. Final sheets were published in color. A series of documents, collectively entitled *The Land of Britain* (published 1937-1947) accompanied the completed maps.

The scope, detail, and speed of Stamp's survey are remarkable, especially in view of the fact that it was completed without the use of aerial photography. The results have formed a set of benchmark data against which scholars and administrators have since

TABLE 3   LEGEND FOR THE LAND UTILIZATION SURVEY OF BRITAIN

Forest and Woodland
  High Forest
    (specified as Coniferous, Deciduous, or Mixed)
  Coppice
  Scrub
  Forest cut and not replanted
Meadowland and Permanent Grass
Arable or Tilled Land, Fallow Land
Heathland, Moorland, Commons, Rough Hill Pasture
Gardens
Land Agriculturally Unproductive, Buildings, Yards, Mines
Ponds

Source: Stamp 1951.

compared existing conditions. During the Second World War, and during the reconstruction that followed, survey results were especially valuable for governmental planners and administrators. As the first modern survey of its kind, the methods developed for the Land Utilization Survey form a precedent and model for subsequent surveys.

Coleman (1961; 1964) described more recent land use surveys in Great Britain, including the Second Land Utilization Survey, introduced in 1961. The second survey is more detailed, in respect to both cartographic and taxonomic detail, but is analogous in many other respects to the original effort.

The Land Classification Program of the TVA

The Tennessee Valley Authority (TVA) was established in 1933 by the U.S. Congress to integrate development of the Tennessee River Basin. The TVA's Land Classification Program was developed and applied under the director of G.D. Hudson (Hudson 1936; Peplies and Keuper 1975; Lounsbury and Aldrich 1979).

TVA's land classification was essentially a land capability categorization that used existing land use as only one of several sources of information. Slope, soil, vegetation, size of land holding, and many other variables were used in a systematic program to assess land capability. The project employed the unit area method for observing and recording land information. In essence, the method is a procedure for systematically coding parcels of land in respect to land use, but also in regard to land capability. The procedure is based upon a concise notation for coding each parcel, as described in detail by Hudson (1936), using a 'fractional code' that records both current land use and physical characteristics of each parcel (Lounsbury *et al.* [1981] provide an example). The unit area method had been developed earlier for recording ground observations, but was applied to aerial mosaics (apparently for the first time) in the TVA surveys (James and Jones 1954). Workers in the field used the mosaics mainly as a base for recording field observations, but it seems inevitable that they must have also used the mosaics for a kind of rudimentary photointerpretation for land use information.

The World Land Use Survey

At the Lisbon Conference of the International Geographical Congress (April 1949) a proposal was made to map the land use of the earth using a common map base and legend. The project was intended as a means of providing land use information useful in the economic development of what is now referred to as the Third World. Then, as now, these nations were often without satisfactory knowledge of their own human and environmental resources. Even the most basic efforts to improve the transport and agricultural infrastructure require knowledge of existing settlement patterns and their relationships to resource distributions (Kellogg and Orvedal 1969). A World Land Use Commission began its work from the premise that ". . . present factual knowledge is inadequate to serve as a proper foundation for schemes of improvement and development, especially in those areas which are commonly regarded as 'underdeveloped' " (IGU 1952:4).

The plan was to map the existing land use of the world, producing descriptive data derived from actual survey, rather than secondary information or subjective assessment. The work was to proceed in several steps (Van Valkenburg 1949; 1950):

(1) *Agreement upon a uniform base map and a common legend.* At a meeting subsequent to the Lisbon Conference, the International Map of the World at 1:1,000,000 was accepted as the standard base for the completed maps. Initial compilation of data was to have proceeded at such levels of detail as required to accurately map land use in each individual region. Although coverage of the International Map of the World was incomplete, sheets of the World Aeronautical Charts, published by the United States Coast and Geodetic Survey at 1:1,000,000, were available for the entire land area of the earth. At the time, this map series formed the largest scale coverage available for the entire earth.

(2) *Staff training in the theory as well as the practice of mapping land use at the proposed scale of the survey.* The proposal explicitly recognized the usefulness of aerial photography in compiling the maps, and acknowledged the requirement for accompanying field observations.

(3) *Actual mapping was to have been completed by individual nations under the supervision of staff trained in the methods and conventions of the world survey.*

(4) *Completed maps were to have been published in a common format, using standardized classification and symbols* (Table 4). All maps were to have been accompanied by descriptive reports.

An international World Land Use Commission appointed by the IGU Congress met in December 1949. The report of the Commission described the classification system, the outline of the project, and the character of preliminary work in several diverse areas of the earth, including India, Cyprus, Switzerland, Iraq, and several regions of Africa. Lounsbury and Aldrich (1979) described work in Puerto Rico that apparently formed one of the pilot studies for the World Land Use Survey.

Considerable effort was devoted to establishing an administrative framework for guiding and coordinating the work. The IGU and UNESCO were involved with the planning from its early days, and numerous national governments and professional societies cooperated with efforts in specific nations and regions. Much of the planning

TABLE 4   LEGEND FOR WORLD LAND USE SURVEY MAPS

1. Settlements and Associated Non-Agricultural Lands
2. Horticulture
3. Tree and Other Perennial Crops
4. Cropland
     a. Continual Rotation Cropping
     b. Land Rotation
5. Improved Permanent Pasture
6. Unimproved Grazing Land
7. Woodlands[a]
     a. Dense
     b. Open
     c. Scrub
     d. Swamp
     e. Cut or Burned Forest
     f. Forest with Subsidiary Cultivation:
          shifting cultivation
          forest crop economy
8. Swamps and Marshes
9. Unproductive Land

---

[a]Woodlands can be further designated as deciduous, broadleaved, evergreen, etc.
Source: Van Valkenburg 1950.

and methodology was based upon experience gained in Stamp's survey of Great Britain. As a member of the Commission, Stamp was personally active in planning the World Land Use Survey and assumed responsibility for coordinating further work in Europe.

A number of publications document the progress of the survey over a period of many years (Tregear 1958; Chistodoulou 1959; Niddrie 1961; Lebon 1965). Coleman (1980) summarized characteristics of major land use surveys, including many of those associated with the World Land Use Survey. Many of these reports differ, it seems, from the objectives of the original concept. They follow a problem-oriented approach to land use study, focusing upon the unique features of each region, rather than the inventory strategy implicit in the original proposals. The two approaches are not necessarily mutually exclusive, but the difference between the original concept and the actual product seems to reveal an absence of focus and control required to coordinate a project of such scope.

A number of reports published as part of the World Land Use Survey appeared piecemeal, without evidence of the administrative and scientific coordination required to implement the original concept. Other documents generated by the Commission outlined the requirement for close coordination, so one assumes that these failings were the result of limited financial support for the project. Although the individual reports have considerable merit, it seems clear that the project as a whole fell far short of its intended goals. Despite the failure to achieve its ambitious objectives, the project did leave a solid body of knowledge regarding the key issues and problems encountered in a broad-scale land use mapping. The requirements for standardization, careful coordination, uniform training of staff, and pilot projects in representative areas of the surveyed regions are clearly documented in Commission reports.

Although few of the published reports are based upon aerial photography, the World Land Use Commission did investigate the possibility of using aerial photographs as a basis for land use mapping. Among the work of the Commission is one of the first formal, systematic investigations of the feasibility of using aerial photography as a basis for land use mapping:

The maps, being compilations from aerial photographs, are experimental, and improvements in technique and accuracy are constantly being made. They have already shown how complex is the pattern in areas previously regarded as relatively uniform. They have shown the need for working on a larger scale than envisaged by the Commission. Obviously the correct procedure will involve spending some time in the field comparing both photographs and maps with conditions seen on the ground. . . . Obviously, too, the interpreters must be experienced geographers, ecologists, botanists or agriculturists, preferably with firsthand knowledge of the areas with which they are dealing (IGU 1952:17).

Marschner's Survey

F.J. Marschner's *Land Use and Its Patterns in the United States* was published in 1959 by the U.S. Department of Agriculture. His work presented a comprehensive description of land use patterns in the coterminous United States and identified physical factors that control the distribution and development of the existing patterns. Marschner organized his work in three main segments. First is a text of about 100 pages describing the existing land use of the United States and historical development of land survey in the United States. Additional sections outline the acquisition of lands by the United States, physical constraints upon land use (including geology, soils, climate, vegetation, hydrology, and industrialization), regional descriptions, and accounts of historical evolution of land use patterns.

The second portion consists of 168 large-scale, black and white aerial photographs representing major land use patterns within primary physiographic and climatic regions in the U.S. Each photograph depicts a pattern, or set of patterns, representing the interactions of cultural processes with the local physical environment — graphic representations of the contrasts between major land use regions. A brief written description identifies the location and key features of patterns on each image.

The third portion is a colored map at 1:5,000,000 reprinted from the 1950 National Atlas Sheet at the same scale. This map was prepared largely from black and white aerial photographs; Marschner reports that he was able to acquire almost complete coverage of the eastern states, but only partial coverage for those in the West. In areas where photographs were not available, or where land use could not be determined from aerial photographs alone, supplementary use was made of pedologic, topographic, and geologic maps. Information from these sources was plotted on state maps at 1:1,000,000. The manuscript maps for each state were used to produce the generalized version published at 1:5,000,000. This national map shows fourteen categories of land use (Table 5). Superimposed over the land use pattern is a dot pattern showing the distribution of cropland as recorded by the Agricultural Census of 1945. For many years, a reduced version of this map formed the basis for the land use

TABLE 5   LEGEND FOR *MAJOR LAND USES OF THE UNITED STATES (1950)*

Cropland and Pasture Land
Cropland, Woodland, and Grazing Land
Irrigated Land
Forest and Woodland Grazed
Forest and Woodland Mostly Ungrazed
Subhumid Grassland and Semiarid Grazing Land
Open Woodland Grazed
Desert Shrubland Grazed
Desert Mostly Ungrazed
Alpine Meadows and Mountain Peaks Above Timberline
Swamp
Marshland
Metropolitan, Cities
Cropland (not irrigated). Includes idle and fallow land and cropland, as
    well as cropland used for pasture (Marschner symbolized this cate-
    gory by a dot pattern superimposed over the solid symbols used for
    the other categories).

Source: Marschner 1959.

map of the United States published in *Goode's World Atlas* (see for example the 12th edition: Espenshade 1960). Later editions have substituted an undated map entitled *Environments* showing nine land cover categories, but with no discussion of source material or method of compilation (see also the map in Thrower 1968). The original map is rather unwieldy and difficult to locate; thus, atlas versions may be more convenient to examine.

Marschner's map illustrates an early use of aerial photography as the basis for a broad-scale land use survey, and forms an example of a map that combines original observations of land cover with secondary census information. It differs from the plan of the World Land Use Survey, and later mapping efforts, in being essentially an individual effort to present a single representation of land use of a very large region at a rather coarse level of detail.

New York's LUNR Survey

The development of New York's Land Use and Natural Resource (LUNR) inventory forms another landmark in the history of land cover and land use mapping. Although it was not the first survey of its kind, it marks an important departure from most earlier surveys and forms a prototype for many of the geographic data bases that have become popular in recent years. In addition, it is notable for its broad areal coverage in relation to the detail and variety of the land characteristics recorded.

The LUNR Survey was developed during the mid- and late-1960s using techniques and concepts conceived at Cornell University for studying the land use of several smaller regions within the State of New York. Many of the methods developed to map these relatively small areas were refined, then applied to the entire state to form the basis of a state-wide information system for land resources. (Hardy 1970; Shelton and Hardy 1974) Some of the important characteristics of the LUNR inventory are:

(1) Aerial photography formed an important source of information for the survey. Its routine use as the basis for a project of this scope forms a noteworthy event in the history of land cover mapping. During and after development of the LUNR inventory, professional journals carried articles discussing the role of aerial photography for land use and land cover mapping, so the LUNR inventory was conducted during a period when scientists were still developing the conceptual and methodological frameworks for use of aerial images in land resource mapping of large areas.

(2) The LUNR inventory recorded, at fine levels of detail, land use/land cover information for the entire land area of New York State. This area is, of course, much smaller than the areas mapped by the programs previously described, but the level of detail is much finer. Greater detail introduced numerous practical problems in the recording, registration and display of information which were encountered for the first time at such a broad scale. The scope of the project required the close coordination of many individuals working as a team.

(3) Use of digital computers to store and manipulate data enabled the survey to record large amounts of data, and efficiently to conduct numerous housekeeping tasks (editing, updating, manipulating, and displaying information) which would be impractical using completely manual methods.

(4) Many of the methods employed for the LUNR inventory can be regarded as rather rudimentary, but they reflect a deliberate effort to develop simple, inexpensive methods that could be easily transferred to other settings. The LUNR survey preceded LANDSAT, mini-computers, widespread availability of digitizing equipment, and the routine availability of high-altitude photography. Therefore, it is important to remember that many technological developments now accepted as commonplace were not routinely available to the LUNR inventory. The survey was specifically designed to use rather simple interpretation equipment and to require only basic skills and experience, an effort to enhance opportunities for transfer to other states and to Third World nations.

(5) The LUNR inventory was perhaps ahead of its time in that it was developed during an era when the merits of land use surveys were not widely recognized. Implementation of the program was approved by Nelson Rockefeller, then Governor of New York, in an act that parallels (probably coincidentally) the Rockefeller Foundation's indirect support of Stamp's 1931-32 inventory of the land resources of Britain. The inventory was eventually discontinued due to its cost and to a lack of interest on the part of those who might have benefited most from the information. It is probably true that the LUNR inventory formed a prototype for

similar surveys in other states — surveys that have continued to provide information of considerable value to land managers and administrators.

Unlike the previous surveys, the LUNR inventory required a uniform spatial unit for data collection and storage, plus an accurate, systematic, geographic reference system to record positions of these units. This requirement followed from the fine spatial detail of the survey and was necessary for convenient revision and updating of information. The LUNR inventory used the UTM (Universal Transverse Mercator) grid system as a locational framework; each UTM cell represents 1 sq km (247.1 acres; 0.381 sq mi) of the earth's surface. Each UTM cell forms a unit of the LUNR inventory. The UTM system forms a worldwide reference system, so it covers the entire state and is compatible with other systems based on the UTM reference.

Land cover and other information for each cell were recorded on computer cards, which were used as the means of storing and updating the information that formed the data base. Manual interpretation of black and white photographs at scales of 1:6,000 to 1:24,000 formed one of the most important sources of information, although pedologic and geologic maps, public land records, direct observation, and other reports and directories were also used. Eleven major categories were used to classify land cover (Table 6). Most categories could be subdivided to distinguish between areal data (recorded as areal proportions of each cell), linear data (presence or absence of highways or railways, for example), and point data (such as the occurrence of buildings within a cell).

The LUNR inventory follows in the tradition of some of the surveys mentioned earlier: land use of a political unit is recorded, then presented in map form. However, the LUNR survey incorporated new features that have subsequently been used routinely in similar systems that cover all or parts of Minnesota, Virginia, and Kansas, for example. These new features include routine use of aerial photography and other remote sensing imagery (often as one of several sources of information), use of computer data bases, provision for routine updates, and the integration of land use data with other physical and socio-economic data.

TABLE 6   NEW YORK'S LAND USE AND NATURAL RESOURCE (LUNR) CLASSIFICATION.

Agriculture
Forest Land
Water Resources
Wetlands
Residential
Commercial and Industrial
Outdoor Recreational Land Use
Extractive Industry
Public and Semi-Public Land Uses
Transportation
Nonproductive Land

Source: Hardy 1970.

USGS Land Use and Land Cover Data and Maps

The United States Geological Survey (USGS) land cover mapping program is an effort to map the land use and land cover of the United States at scales of 1:250,000 and 1:100,000. Patterns of land use and land cover are classified using the system designed by Anderson and colleagues (1976) specifically for mapping at the rather broad scales necessary for a nationwide survey. James Anderson and Richard Witmer at the USGS collaborated with Ernest Hardy and John Roach at Cornell, who were able to contribute experience acquired during development of the LUNR inventory. The classification system is designed specifically for use with remotely sensed data, but was developed after careful examination of existing land use classification systems in use throughout the nation, as well as existing maps and data (Table 7). Its hierarchial structure, with two published levels of detail, permits convenient use with remotely sensed data at varied scales and resolutions. Levels I and II are appropriate for mapping rather coarse levels of detail, such as the scales of 1:250,000 and 1:100,000 used in the USGS program, but are compatible with more detailed classifications at level III that can be developed for local use. Whereas levels I and II are generally applicable within the U.S. and Canada, definitions of appropriate level III categories will vary depending upon the local setting, the purpose of the survey, and the kinds of imagery available.

The USGS program records land use and land cover data (Figure 4) at scales of 1:250,000, using the USGS one degree by two degree 1:250,000 map series as a base (Anderson 1977). As maps at 1:100,000 become available they will form the base for land use data, although the information will retain the same level of detail shown on the 1:250,000 sheets. Land use and land cover information is interpreted from high altitude black and white, color and color infrared positive transparencies, then recorded directly on scribecoat masters that form the basis for subsequent maps and data. Interpreters use standard equipment and the usual monoscopic or stereoscopic interpretation methods assisted by binocular microscopes.

The USGS program generates at least six forms of data that must be accurately registered to each other in both cartographic and digital form (of course, only the land use data are interpreted from aerial photography):

(1) Land use and land cover data,
(2) Boundaries of political units,
(3) Hydrologic information,
(4) Census county subdivisions,
(5) Federal land ownership, and
(6) State land ownership.

A completed land use/land cover map sheet could include as many as 37 separate categories. A full sheet might have 4000 polygons, with some 10,000 separate line segments. Complete coverage of the United States (on the one by two degree sheets of the 1:250,000 series) consists of 631 sheets.

In part because of the immense task of generating, storing, manipulating, editing, and analysing so much information, the USGS created the GIRAS (Geographic Information Retrieval and Analysis System) for manipulating the data in digital form (Mitchell et al. 1977). GIRAS permits convenient handling of digitized land use information. Although the exact details of the system have varied as it evolved, it is a batch system

TABLE 7   THE 1976 USGS LAND USE AND LAND COVER CLASSIFICATION

Level I		Level II	
1	Urban or Built-up Land	11	Residential
		12	Commercial and Services
		13	Industrial
		14	Transportation, Communications, and Utilities
		15	Industrial and Commercial Complexes
		16	Mixed Urban or Built-up Land
		17	Other Urban or Built-up Land
2	Agricultural Land	21	Cropland and Pasture
		22	Orchards, Groves, Vineyards, Nurseries, and Ornamental Horticultural Areas
		23	Confined Feeding Operations
		24	Other Agricultural Land
3	Rangeland	31	Herbaceous Rangeland
		32	Shrub and Brush Rangeland
		33	Mixed Rangeland
4	Forest Land	41	Deciduous Forest Land
		42	Evergreen Forest Land
		43	Mixed Forest Land
5	Water	51	Streams and Canals
		52	Lakes
		53	Reservoirs
		54	Bays and Estuaries
6	Wetland	61	Forested Wetland
		62	Nonforested Wetland
7	Barren Land	71	Dry Salt Flats
		72	Beaches
		73	Sandy Areas other than Beaches
		74	Bare Exposed Rock
		75	Strip Mines, Quarries, and Gravel Pits
		76	Transitional Areas
		77	Mixed Barren Land
8	Tundra	81	Shrub and Brush Tundra
		82	Herbaceous Tundra
		83	Bare Ground Tundra
		84	Wet Tundra
		85	Mixed Tundra
9	Perennial Snow or Ice	91	Perennial Snowfields
		92	Glaciers

Source: Anderson *et al.* 1976.

with an organization providing capabilities for data input and editing, data retrieval and manipulation, and data display and output.

Data input is by digitization — the process that converts the graphic information scribed by the interpreters into digital form subject to manipulation by computer. This process is among the most complex and the most important of any in the system, as it determines the overall efficiency and accuracy of the graphic display. Initially the USGS experimented with several approaches to digitization, including manually operated cursors, before depending mainly upon use of a laser scanner. Data input is closely associated with editing as only clean, accurate data can be passed on to subsequent steps in the process. As a result, data are presented in a standard format, edited for errors, and reduced in volume (for example, by removing non-essential points in line data).

To permit efficient retrieval and manipulation, the GIRAS structure was designed specifically, using accepted principles for handling spatial data, to handle geographic information in polygon form. Separate files for each map section retain information concerning arc records, coordinate data, polygon data, and information relating specific arcs to polygons. Information can be retrieved by area, or by selecting features or classes of categories. Full maps or portions of map sheets can be displayed using drum plotters, CRTs, line printers, or other devices. Masters for color maps can be generated using peelcoat cut by flatbed plotter.

The boundaries of political and census units, as well as major drainage basins have also been digitized so that land use and land cover data can be retrieved, displayed, and tabulated in a format compatible with their intended use and with other data. As a consequence, it will be possible to associate changes in land use over time, for example, with variations in stream discharge data routinely collected by the USGS, and with other demographic and agricultural data gathered by other governmental agencies.

Although land cover data are gathered at levels of detail generally considered far too coarse for effective use at local levels of government, pilot studies have demonstrated their utility at state and regional levels. Completion of coverage for the entire nation will permit examination of relationships between physical and demographic issues that would not have been previously possible at national scales. Especially intriguing are the opportunities presented by repetitive photographic coverage for updating the nation's land use and land cover information. Planned updates of land cover information present an opportunity to systematically monitor changes in land cover over time, and thereby to assist in dealing with long term problems such as losses of agricultural land to non-agricultural uses or understanding the hydrological implications of demographic, economic, and land use changes within specific drainage basins.

Aerial imagery is now used in the completion of tasks formerly undertaken by direct observation from the ground. If we compare many of the most recent land cover inventories with those produced before the use of aerial imagery, we find few real innovations in procedure or form. Instead the formats accepted (for example, by Stamp's inventory) survive today with only minor modification in current maps and inventories. As a result, most of the conventions that we accept as standard practice for the preparation of land use surveys have origins in earlier practices. Most of the inventories described present land cover patterns on map bases derived from standard topographic series or their equivalent. Land use information is therefore presented in clear relationship to cultural and physiographic landmarks known to the reader or identifiable from readily accessible sources. Use of an accurate planimetric base permits measurement of areas and representation of correct spatial position.

With respect to classification systems, these inventories exhibit some consistency in regard to identities of categories, but there has been some change in logical organization. Many of the earlier systems (such as those of Stamp, Marschner, and the LUNR survey) are organized at a single level of detail, although selected categories are favored with subdivisions. During the interval that some of these classifications were in use, activity codes designed for ground survey at the local level were routinely organized in a hierarchial manner using several levels of detail. As a result, it is surprising to find that it was not until the 1970s that fully hierarchial classifications were formally proposed for use with remotely sensed data (Anderson 1971; Anderson et al. 1976).

In the use of aerial imagery, the inventories described here cover a spectrum ranging from no use whatsoever to significant dependence. Stamp's survey relied upon direct observation in the field, with no use of aerial photography. The TVA survey used aerial photography as a kind of cartographic base for recording field observations in a region that was without accurate base maps at the time. No doubt photographs were also used as a rudimentary source of information in inaccessible regions. Marschner's survey used aerial photography as a means of delineating broad-scale land use regions, but did not attempt to map at the fine level of detail used by Stamp and others. Methodologically, if not chronologically, Marschner's survey represents an intermediate phase in the use of aerial imagery as a source of information for detailed land use inventory. The pilot studies for the World Land Use Survey applied aerial photography as a source of land use information at rather fine levels of detail, and developed the procedural and conceptual frameworks for routine use of aerial imagery for land use mapping. Finally, the land use inventories conducted by the LUNR staff and the USGS have developed comprehensive land use inventories based almost exclusively upon manual interpretation of aerial imagery.

Estes (1982:27) noted that despite the broad scope of research in remote sensing, "The only really proven technique is conventional photo interpretation." As a result, it seems clear that we have yet to experience the real influence of recent advances in remote sensing technology upon the forms, conventions, standards, and methods that we now accept as standard for land use inventory. As users of remote sensing data come to use an increasing proportion of the available remotely sensed data, we are likely to see changes in the symbols, formats, and conventions used to prepare land cover maps. It is now too early to suggest what these changes may be, but we can expect them to evolve from rather than replace practices currently accepted as standard for manual interpretation of aerial images.

FIGURE 4   PORTION OF A USGS LAND USE AND LAND COVER MAP. This map shows land use in the Harrisburg, PA region; it registers to the Harrisburg, PA. 1:250,000 topographic map published by the USGS. Here land use is shown in the conventional format — polygons are designated by numbers that correspond to categories in a classification system. Here the numerical symbols designate level II categories in the USGS system (Table 7). These symbols, of course, identify only the predominant land use category for each polygon. This map (also shown in color on the cover of this volume) is typical of those in the series produced by the USGS for the United States. Land use information is interpreted from aerial photography, then portrayed at 1:250,000 as illustrated here. This includes a complex mixture of forested ("41") and agricultural land ("21") together with urban and suburban land ("11," "12," "16," "17") associated with built up areas in and near Harrisburg. The Susquehanna River flows from northwest to southeast across the area; also visible is a portion of the transportation network formed by the junction of several interstate highways. (Map provided by the Geography Program, U.S. Geological Survey).

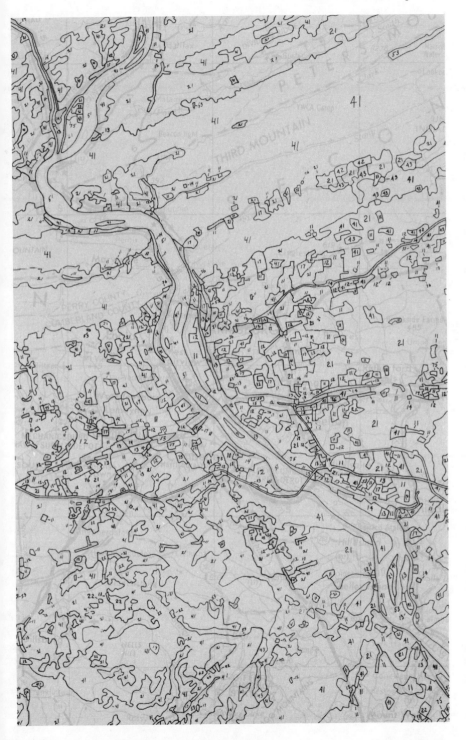

4

Principles and Conventions for Preparing Land Use Maps

The earth's land use and land cover patterns are symbolized on maps by a mosaic of discrete parcels, each assigned to one of several land categories. Land use maps therefore use a cartographic model that requires every location in the mapped region be assigned to a single land use category, such as agricultural, urban, or forested land. Land use categories actually differ from those used in many other mosaic maps of similar appearance because they are defined solely by qualitative distinctions between categories, whereas categories in many other maps of similar appearance (such as pedologic, climatic, or geologic maps) may be based in part upon quantitative differences between categories. In most respects, however, land use/land cover maps can be regarded as typical members of the class often referred to as mosaic maps, patch maps, or "chorochromatic" maps (Monkhouse and Wilkinson 1971).

As members of this cartographic family, land use/land cover maps use conventions of logic, symbolization, and organization common to maps of the class. They are, of course, subject to the same errors and limitations. To effectively make and use these maps it is important to understand common conventions and their limitations.

A literal interpretation of the cartographic model just described would translate each parcel on the map to single corresponding discrete and homogeneous parcel on the ground. Seldom is this ideal encountered in practice. Actual boundaries between parcels may not correspond to the abrupt lines used to symbolize edges of parcels. Boundaries may sometimes be diffuse transition zones of considerable width. Homogeneous parcels visible on the map may actually encompass inclusions of foreign (not symbolized) categories too small to depict at the scale of the map. Seldom can the image interpreter completely avoid errors in the identification of categories or placement of boundaries.

Some of these errors can be avoided, minimized, or corrected by acceptance of high standards in the preparation and editing of manuscript maps. Other errors are the inevitable artifacts of complexity of the landscape and characteristics of the cartographic models that we use. Especially bothersome are those errors caused by the presence of several categories within parcels that must be represented as a single category, due to the inability of the cartographic model to portray all of the variability present in the landscape. Such errors are the unavoidable consequence of mapping land use, yet they influence map accuracy and credibility, especially in the eyes of readers inexperienced in the use of maps.

As a result, the image interpreter must devote special attention to preparing accurate descriptions of the actual identities of categories (as distinguished from the idealized nominal identities). Thus, a category designated as agricultural land may include patches of forested land too small to delineate legibly at the scale of the map. The category description must be written to inform the reader that significant impurities are present within parcels labeled as agricultural land. Thus, it is possible to inform the reader of presence of impurities within certain categories on the map by presenting a written description of each category describing the existence, patterns, and characteristics of inclusions.

Several concepts implying quality should be defined explicitly. *Accuracy* refers to correct assignment of parcels to categories. Urban land mapped as urban land is "accurate"; urban land mapped as agricultural land is "inaccurate." Note that accuracy in itself is not always a satisfactory measure of the usefulness of a map for a specific purpose. For example, if we categorize land as either 'water' or 'land,' we can often attain very high accuracy, but the results may not be useful for practical purposes because the level of detail is so low. We prefer usually to have greater *precision,* which can be defined as the measure of detail in a map or classification system. The land/water separation therefore may permit attainment of a very high accuracy, but very low precision. For some purposes, low precision may be perfectly satisfactory for the purpose at hand, although we usually desire at least some modest level of detail in land use classification and mapping. To attain greater precision, more categories are added to the classification and finer spatial detail is represented on the map. In many cases, it appears that increasing the number of categories also increases the opportunity for errors in boundary placement and parcel identification. Thus, there is a tradeoff between accuracy and precision. As we attempt to show finer detail, so too we tend to make more errors.

One of the most difficult practical problems in preparing land cover maps from remotely sensed data is selection of appropriate levels of precision. As more categories are added to the classification system in an effort to improve precision, the numbers of parcels represented on the map becomes larger, and their sizes become smaller. The increased detail in the classification system can be designated as 'taxonomic' detail, or *taxonomic precision,* while the detail in the map pattern (i.e., the sizes of the parcels on the map) can be referred to as 'spatial' detail, or *spatial precision.*

As both forms of precision increase, the visual and local complexity of the map increases, along with opportunities for errors. The correct balance between complexity and precision must be defined by the image interpreter, who must consider the user's requirementts for detail in relation to the natural complexity of the landscape,s represented on the image.

Principles of Land Use Mapping

It is impossible, even if it were desirable, to specify beforehand the exact characteristics of every map, so each map maker must make innumerable decisions regarding map form and content. We prefer generally that each map maker should not make these decisions at random, or solely according to his or her individual idiocyncracies, but rather by following conventions and principles accepted by the larger community of map makers and map readers. For the most part, these principles and conventions seem to be rather informal and implicitly rather than formally expressed.

Although no single map is likely to be perfect in all respects, there is merit in proposing an explicit listing of desirable qualities for land cover maps, as they provide a means of assessing the quality of specific maps. Because so many of these principles are implicitly rather than explicitly expressed in the work of experienced geographers, it seems useful to express formally some of these principles for the benefit of students who may be uninitiated in respect to standards for preparing maps and data.

Legibility

Each map should be clear and easily readable at publication scale. Symbols must be concise and easily distinguished. The sizes of the smallest parcels must be selected so that they are legible, and can be labeled with symbols. Often this requirement means that very small or very narrow parcels must be omitted, or possibly exaggerated in relative size, so that they will be legible.

Sizes of parcels depicted on the map are, of course, a function of the mosaic of parcels on the landscape as they interact with the detail of the classification system. Therefore, each map must be designed with consideration of the interplay between spatial and taxonomic detail. The manuscript map must be prepared with knowledge of the final publication scale and the method of reproduction, so that the final map, as well as the manuscript map, meets requirements for legibility.

Accuracy

The reader of a map has no choice except to assume each parcel on the map possesses its stated identity, unless otherwise informed. In areas of complex land use patterns, completely accurate delineation of uniform parcels may be impractical. Therefore, in such instances the reader must be informed of the presence of impurities within nominally uniform categories by means of statements in the legend or in the report that accompanies each map. Complex mapping units can be explicitly described as 'transition zones,' or 'mosaics' as outlined in Chapter 5, or by identification of such parcels at more general levels in the classification (to avoid the errors that accompany specificity).

Explicit Description

The reader of a map should be able to learn all relevant information concerning the meaning of the map and the method of compilation from information presented on the map itself, its legend, and the accompanying report. Source materials should be specified explicitly; remotely sensed data or imagery should be specified by date, scale, type, quality, and format. If several sources of imagery have been used, each should be described using a small-scale coverage diagram (Figure 5) to represent comparative coverages. Each category must be explicitly described in the report, including specific accounts of category definitions, enumeration of component features, and clear descriptions of the image appearance of each category. Descriptions must explicitly outline the composition of mapping units characterized by mixtures of several categories, or any other departures from the nominal identities of mapping units.

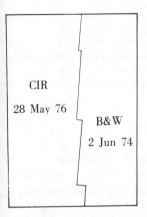

FIGURE 5 EXAMPLE OF A COVERAGE DIAGRAM. The sketch map indicates coverage of two kinds of photography within the boundaries of the mapped area. The coverage diagram is, of course, not required if the entire map is represented on a single form of imagery. In this example half of the mapped area was covered by color infrared photography (CIR); the remainder is covered by black and white photography (B&W) acquired on a different date.

Consistency and Uniformity

Each map maker should aspire to present uniform, consistent detail throughout the map. Each parcel within a given category should represent (in the ideal) geographic areas of uniform land use, or relatively uniform mixtures of land uses, as described in the legend and report. The reader should be able to examine a map and find that the true identities of parcels (as observed on the ground) consistently match the cartographic representation, regardless of placement within the mapped region. Ideally, accuracy should not vary from one portion of the map to another. Furthermore, mapping detail must be consistent throughout the mapped region. The reader should be able to examine a map with the understanding that variations in the sizes and the identities of parcels represented on the map reflect genuine variations on the ground.

Accurate Planimetric Representation

Land cover/land use detail must be presented to the user on an accurate planimetric base. Information derived directly from the image itself is not suitable for accurate measurement of areas and distances — operations basic to the use of any large scale map — due to the inevitable presence of geometric errors in all forms of remotely sensed data. As a result, land cover information must be presented in correct planimetric form, as described later. This requirement means that the interpretation of information from an image must proceed in two separate steps. First, information is interpreted directly from the image, usually on a transparent overlay that registers directly to the image *(the image overlay)*. Second, information from the completed image overlay is plotted (with appropriate changes in scale and geometry) on an overlay that registers to an accurate map *(the map overlay).* The map overlay forms the basis for the final map that is presented to the user, and is the basis for all areal measurements required for the study. Usually the image overlay is of significance only to the interpreter, who retains it as a working document in the event it is necessary to check or revise the original interpretation.

Compatibility with Other Data

Land cover/land use information has its greatest usefulness when it is compatible with other information in respect to time of collection, level of spatial detail, units of collection, and taxonomic detail. Although it is unlikely that a map maker could expect to control all of these aspects of his project, it may be possible to choose between (for example) imagery acquired at several dates, or to aggregate data by political units within the study area.

The concept of compatibility applies in several respects. It might be desirable, for example, to prepare land use data in a manner that promotes compatibility with census data in respect to detail and timing. A survey designed to monitor land cover changes should consider the dates and scales of previous surveys, as well as the structure and detail of the earlier classification.

Appropriate Taxonomic and Spatial Detail

Implicit throughout this discussion is the concept that the maps are not prepared in isolation from a knowledge of their ultimate uses. Those who prepare the maps and data must be informed of the requirements of the individuals and organizations who will use the information. In specific, the detail and the organization of the classification, as well as the spatial detail of the final map, are usually the key characteristics that determine usefulness for specific purposes. (Other items of significance may include the date of the information, compatibility with other data, and the choice of map base.) Usually those compiling the data should first coordinate their efforts with those who will use the data before the survey is initiated, again as the image overlay is completed, then once more just before final versions of the maps and reports are prepared.

Form and Symbolization on Land Use Maps

Apparently, the overall form of land cover/land use maps and the symbolization used have changed little since their development in the 1930s. Although exact methods and procedures of compilation have varied considerably, current maps are visually and logically very similar to those produced (for example) by Stamp's survey during the 1930s. Some of the key characteristics of current land use maps can be enumerated:

(1) Use of a planimetrically accurate base permits preservation of correct geographic position and the measurement of correct areas and distances. Stamp's survey used a large scale topographic map series as a base, a practice that has been followed in later projects, including the USGS program. Often the land use symbols can be superimposed over selected hydrographic, political, or topographic data to provide a convenient locational reference for the reader.

(2) Categories are identified on the map by symbols (usually letters or numerals) keyed to categories in a classification system (Figure 4). Typically maps are prepared using solid lines to separate parcels of separate categories; the reader must interpret each parcel on the map as a uniform region of land use. For small scale maps, especially, this interpretation may seldom be justified, so the reader should be informed of the existence and character of inclusions within mapping units.

(3) For many purposes, black and white maps are the most practical form for cartographic representation of land use data. Manuscript and printed maps can both be prepared quickly and inexpensively without the problems encountered in color reproduction. Such a map consists primarily of the outlines of land cover parcels, with symbols placed inside each parcel. Symbols consist of one to three digit numerals corresponding to the categories in the Anderson system or its equivalent. Many other classification and symbolization strategies are possible, but it seems sensible to accept the Anderson system as a concise, efficient standard in the absence of reasons to adopt another system.

(4) Often the land use/land cover map is prepared as an overlay to a general purpose map of the region of interest. Because the overlay is prepared on a transparent base to permit the land use distribution to be viewed in relation to cultural and topographic features, the use of colored patterns is inappropriate. They obscure detail on the base map. In other situations colored symbols may be appropriate, especially if the map is to be used before large groups at meetings who may not be able to examine the map in detail. Several systems for color symbolization have been proposed (Table 8).

(5) Classifications differ greatly in definitions of categories, numbers of categories, and logical structure. The use of a hierarchical classification is especially important for applications to land cover mapping because imagery at varied scales and resolutions may be used; hierarchy also permits generalization of the data at different levels of detail.

(6) Details of marginal information on maps has varied. Usually the title should identify the geographic location, the nature of the informational content, and the date of the information. (For example: "Land Use and Land Cover of the Roanoke, VA Area, May 1979.") A bar scale is preferable to a word statement of scale or a representative fraction; it may later be necessary to change map scale photographically. Each map should, if appropriate, include a coverage diagram as a key to the varied imagery used to compile a specific map (Figure 5). Usually it is appropriate to include information that permits the reader to establish location within a geographic reference system, such as latitude and longitude, or the UTM grid. A simple locational diagram (Figure 6) is similarly useful.

TABLE 8   CATEGORIES AND SYMBOLS IN THREE LAND USE SURVEYS

Category	Symbol	Color
Land Utilization Survey of Britain (1931-1938)		
Forest and Woodland	**F**	Dark Green
Meadowland	**M**	Light Green
Arable Land	**A**	Brown
Heathland	**H**	Yellow
Gardens	**G**	Purple
Agriculturally Nonproductive	**W**	Red
Ponds	**P**	Blue
World Land Use Survey (1952)		
Settlements and Associated Non-Agricultural Lands	**1**	Red
Horticulture	**2**	Deep Purple
Tree and Other Perennial Crops	**3**	Light Purple
Cropland	**4**	Brown
Continual and rotation cropping	**4a**	Dark Brown
Land rotation	**4b**	Light Brown
Improved Permanent Pasture	**5**	Light Green
Unimproved Grazing Land	**6**	Orange, Yellow
Woodlands	**7**	Green
Swamps and Marshes (nonforested)	**8**	Blue
Unproductive Land	**9**	Gray
Anderson/USGS Level I Categories (1976)		
Urban or Built-Up Land	**1**	Red
Agricultural Land	**2**	Light Brown
Rangeland	**3**	Light Orange
Forest Land	**4**	Green
Water	**5**	Dark Blue
Wetland	**6**	Light Blue
Barren Land	**7**	Gray
Tundra	**8**	Green-Gray
Perennial Snow or Ice	**9**	White

(7) Seldom can a map stand by itself without some form of accompanying report that describes the map and its preparation. Such a report should provide four functions: (a) explanation of how the map was prepared, including identification of the source materials, (b) regional description of the area represented on the map, at least in rather broad terms, so the reader has a basis for relating land use patterns to the physical, cultural, and economic setting of the region; (c) a summary of the essential features of the observed land use pattern, including tabulation of the areas occupied by each category; and (d) clear, precise descriptions of categories used on the map. Preparation of this report is described in the next chapter.

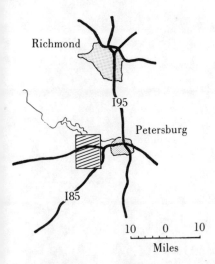

FIGURE 6 EXAMPLE OF A LOCATIONAL DIAGRAM. This sketch map shows the location of the mapped area (the shaded pattern) in relation to distinctive drainage, transportation, and settlement features.

5

Manual Interpretation
for Land Use Mapping

Image interpretation has been defined as ". . . the act of examining images for the purpose of identifying objects and judging their significance. Interpreters study remotely sensed data and attempt through logical processes to detect, identify, measure and evaluate the significance of environmental and cultural objects, patterns, and spatial relationships" (Estes and Simonett 1975:869). Interpreters use *elements of image interpretation* to detect, recognize, and identify objects and patterns. These elements traditionally include size, shape, tone, texture, pattern, and association — that is, those qualities that permit us to recognize features we see on aerial images. Although we routinely use these characteristics in everyday life to recognize objects, we must formalize the interpretation process to apply it in the more abstract and unfamiliar context of remote sensing.

Use of aerial imagery for mapping land cover calls for application of skills not normally required in the simple, intuitive examination of images for recognition of individual objects. Individual land cover categories are formed from collections of diverse objects, features, and structures that are often not individually resolved on the image; the interpreter's task is not so much one of identifying separate objects as it is the accurate delineation of regions of relatively uniform composition and appearance. The interpreter of land use information must, then, generalize to define the areal units that compose the subject of the interpretation. The goal should be to perform this mental generalization in a consistent, logical manner and to describe the procedures accurately in a written account of the process of manual interpretation of images, and their application to land use mapping.

Image and Interpretation

All remotely sensed data can be represented either in *pictorial* form or in *numerical* form. In black and white image format, data appear as tones of black, gray, and white. At a fine level of detail, an image can be resolved into individual elements, often too small to be recognized by the naked eye. In many images, these individual elements form uniform subdivisions similar to cells in a uniform grid, referred to as picture elements, or *pixels,* that represent the brightnesses of very small portions of the scene. By examining an image at a level of detail fine enough to separate these individual elements, we can represent an image as an array of values, with each value corresponding to the brightness of a single pixel. (Each pixel in turn represents the reflectance of a specific area on the earth's surface.) Depending upon the characteristics of specific sensors, images may take their original form as black and white images, or as arrays of numerical values. Equipment is available to change images to arrays of values (optical digitizer) and arrays of values to images (video display). The most desirable format depends upon the purpose of the investigation, the remote sensing system to be used, and the resources available to the image analyst.

Numerical Interpretation

An examination of a numerical representation of a scene entails statistical and mathematical manipulation and analysis of the values that represent reflectances. The numerical approach to remote sensing can be described as quantitative, abstract, and dependent upon automated or semi-automated techniques. Once equipment and software are on hand, it may be cheaper and faster, for large amounts of data, to use the numerical approach than to perform the same interpretation manually. The numerical approach often has advantages for operations that must be repeated with consistency, and for interpretations that must simultaneously examine data in several portions of the spectrum.

Numerical interpretations usually rely extensively upon spectral information (brightness measured in several portions of the spectrum). In the ideal, features on the earth's surface may be said to display distinctive patterns of spectral reflectance (sometimes referred to as "spectral signatures") that permit specific crops, soils, or land cover categories to be recognized. The digital format is especially powerful in this context because it permits a simultaneous examination of several spectral channels, the use of statistical concepts in forming decision rules, and the application of other strategies not practical in manual interpretations. Most algorithms for numerical interpretation, however, are not yet able to routinely exploit the spatial, or textural, information contained in the relationships between brightnesses of neighboring pixels. Because image texture is the source of much of the ability of human interpreters to accurately interpret complex scenes, the inability of numerical procedures to completely duplicate human capabilities is an important limitation in the application of the numerical approach in the classification of land cover and land use.

Manual Interpretation

The examination of an image as a print or transparency using the elements of image interpretation is referred to as 'manual interpretation.' This procedure can be characterized as *traditional, concrete,* and *qualitative* in nature. An interpreter uses brightness information (tone), as does the quantitative approach, but also makes extensive use of spatial information in the image. The interpreter does not, of course, examine an image pixel by pixel, to evaluate each pixel in isolation, but instead looks at the brightness of each pixel in relation to its neighbors. It is the use of this spatial information that permits human interpreters to make the complex identifications and analyses that are possible in remote sensing. Although automated interpretations make some use of spatial information, it is usually at a very primitive level. On the other hand, most human interpreters can make only limited use of the spectral information so effectively used in automated interpretation. For manual interpretation, the focus is upon delineation of *areas* of relative uniformity at the working scale and resolution. The identification of *objects,* although of obvious significance, is of somewhat lesser importance.

Quality of Manual Interpretation

Image interpretation seems to be a skill that is in part a natural aptitude, and in part an acquired skill that can be enhanced by experience, study, and training. A proficient image interpreter possesses a good background in the specific field of knowledge at hand, as well as a thorough knowledge of principles of remote sensing and the specific geographic region under examination.

The beginning student of image interpretation must make several adjustments to everyday experience, and develop new forms of intuition and experience. First is a *difference in perspective.* We are accustomed to examination of objects from a perspective near ground level. On aerial imagery we see the same features from above, and must adjust to new relationships between illumination and shadow, and see portions of objects not often visible in our normal experience. In everyday experience, a meadow may appear as a lush, green area because at ground level we see mainly the sides of plants. From the overhead perspective of aerial imagery, we may find that the same field appears as sparsely vegetated because we see a greater proportion of bare soil between the individual plants.

Second, remotely sensed imagery often uses *energy outside the visible portion of the spectrum.* Many of the implicit rules of visual recognition derived from everyday experience no longer apply in other portions of the spectrum. Evergreen forest frequently appears to be dark in the visible spectrum, but is very bright in the near infrared. A gravel road reflects as a rough surface in the visible spectrum, but may appear as a smooth surface to microwave radiation.

Another difference in the representations of features on aerial imagery is due to unfamiliar levels of spatial and radiometric resolution. *Spatial resolution* refers to the ability of the image to legibly record small objects. Remotely sensed images, by necessity, represent landscapes at coarser resolutions than we are accustomed to in everyday experience. Frequently at coarse spatial resolutions familiar objects may be unrecognizable unless the interpreter is mentally prepared for the differences. *Radiometric resolution* refers to the ability of a sensor to separate varied degrees of brightness. If radiometric resolution is low, a scene may be represented mainly in whites and blacks rather than in the range of grays that may permit the interpreter to distinguish greater detail.

Image scale is a characteristic quite separate from spatial resolution; scale can be changed independently of resolution. We can adjust to simple changes in image scale easily if other qualities have been held constant. More difficult for the beginning interpreter are changes in *geometry,* the positional relationships between features on the image. Each form of remote sensing imagery represents positions of features on the earth's surface in characteristic geometric relationships that differ from their true relationships on the ground. For example, terrain slope and elevation may be portrayed on side-looking airborne radar imagery in a manner that includes systematic positional errors. Relief displacement is a major source of positional error in aerial photographs that causes misrepresentation of the locations of elevated features in regions of uneven topography. Data from the Landsat multispectral scanner includes a variety of systematic positional errors. Every image interpreter must acknowledge the existence of geometric errors and assure that final interpretations are presented in a planimetrically correct format.

The degree to which specific objects or boundaries are visible on an image is a function of the spectral, spatial, and radiometric resolutions of the sensor system, the atmospheric conditions at the time of image acquisition, and the character of the landscape imaged. Landscape variables include the *sizes* and *shapes* of land cover parcels, which obviously influence the recognizability of specific categories. Also important is the *contrast in brightness* between adjacent parcels. The internal uniformity and the sharpness of boundaries are also likely to be important. In general, one would expect large, uniform parcels with distinct edges, regular shapes, and high contrast in brightness with neighboring parcels to be easiest to delineate. The fact that parcels of a given category possess these qualities to varying degrees even on a single image may explain why accuracy of delineation can vary even for a single category.

Elements of Image Interpretation

Although image interpretation is an extremely complex process that we do not really understand in detail, it is possible to identify some of the key attributes of image representation that enable us to recognize and identify features depicted on images: size, shape, tone, texture, shadow, pattern, and association (Avery 1977; Colwell 1960). In image interpretation, we must make a disciplined effort to use these same elements to analyze image representations of objects and areas. Knowledge of the elements of image interpretation serves two purposes. First, conscious application of these elements may permit identifications that would not otherwise be obvious. Second, these elements provide a vocabulary common to image interpreters that permits concise description of the image appearance of specific features.

These elements are applicable equally well to imagery collected by all remote sensing systems (Lillesand and Kiefer 1979; Bryan 1982), although in each instance the interpreter must acquire, then apply, a thorough knowledge of the specific sensor and its imagery.

Size.   The sizes of objects are among the most useful clues to their identification. For land cover mapping interpretations, it is seldom necessary to make detailed measurements of size for the purpose of identification, but it is clear that the sizes of land cover parcels are often among their most distinctive characteristics.

Shape.   The shape of the outline of a land cover parcel can also be a distinctive characteristic. For example, cropped agricultural land frequently appears in regularly shaped parcels, whereas pasture may tend to occur in parcels of irregular or indistinct shape.

Shadow.   Shadows are traditionally one of the most important factors in the identification of objects on aerial photographs because they can reveal the silhouettes of objects otherwise visible only from above. For interpretation of land cover, shadows assume a somewhat different, but equally important, role. We usually are not interested in the identification of specific objects, but rather in the identification of areas. Shadows can be important in this context by enhancing boundaries between categories, and in contributing to textural differences. For example, the contact between an open field and an area of mature forest, if illuminated from the forest side, will be marked by the shadows of the forest canopy as it falls on the field (Figure 7). As viewed from above, the shadow forms a dark line at the edge of the field that enhances the boundary between

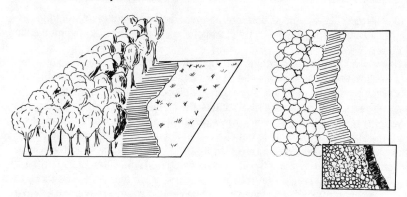

FIGURE 7 THE SIGNIFICANCE OF SHADOW IN LAND USE INTERPRETATION (I). The sharp contact of forest and pasture permits the shadows to fall on the open land (left hand diagram). As viewed from above (right hand diagram), the shadow forms a dark strip that parallels the boundary between the two parcels. The inset represents a small scale view illustrating how the shadow is visible as a dark ribbon that enhances this segment of the boundary.

the two categories. A second example (Figure 8) illustrates the contribution of shadow to the textural differences between categories. An open field partially vegetated by widely spaced saplings will display a distinctive image appearance. As seen from above, the crowns of the saplings form isolated, round, dark spots; the shadows of the saplings fall on the grassed areas. The interpreter sees this pattern on the image as a speckled region, that, once recognized, forms a distinctive signature for finding similar

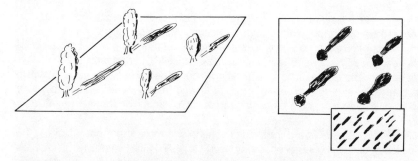

FIGURE 8 THE SIGNIFICANCE OF SHADOW IN LAND USE INTERPRETATION (II). The left hand diagram represents small trees and shrubs positioned in an open pasture. Their spacing is wide enough to permit their shadows to fall on the open grassland between trees. When such areas are viewed from above (right), the interpreter sees only the crowns of the trees, and their shadows. The inset shows an overhead view at small scale; the parallel streaks of the shadows create a smeared appearance that forms a distinctive signature for this kind of land cover.

areas elsewhere on the image. Once the interpreter explicitly recognizes the contribution of shadowing to the appearances of land cover categories, it becomes one of the most important clues to mapping land cover.

Texture.   Texture is the distinctive visual impression of roughness or smoothness, caused by variability or uniformity of image tone, characteristic of certain features and areas as represented on aerial images (Figure 9). By definition, texture must apply to areas of land (rather than to individual objects), so it forms another useful property for identification of land cover. The character of forested areas can, for example, often be related to image texture. Smooth textures are commonly associated with young stands of trees; rougher textures usually indicate more mature trees with fully developed crowns.

Image texture seems ultimately to be related to variations in the heights of surfaces and objects within the area of interest. Thus, areas characterized by complex assemblages of objects and structures usually have rather rough, uneven textures (assuming that the sun elevation is rather low). Suburban scenes, for example, formed by groupings of detached dwellings, large trees, grassed areas, and pavements, often appear as rather rough textures. Large cropped areas, with plants all at the same height, typically appear as smooth textures.

Tone.   Image tone can be defined as a distinguishable shade of gray from white to black. Tone probably provides more information than any other single element of image interpretation. Contrast in tone is of course what permits the representation of features on the image. Typically land cover parcels are delineated by outlining areas that have more or less uniform image tone (although often image texture may assume great significance in delineating land cover parcels).

FIGURE 9   CONTRASTING TEXTURES. This Virginia scene is composed of open water, forest, highways, and grassland, all identifiable primarily by differences in tone and texture. Note the role of shadow in determining texture.

FIGURE 10   PATTERN. An arrangement of parcels in a pattern characteristic of cropped agricultural land in Arizona (X-band radar image courtesy of Goodyear Aerospace Corporation and the U.S. Air Force).

Pattern.   Pattern refers to the overall form of an array, or series, of related objects. A classic example is the appearance of an orchard on aerial imagery; the regular spacing and placement of trees forms the distinctive image pattern easily recognized by all interpreters. The orchard forms an example of pattern within a land cover parcel. Of wider significance is patterns of occurrence between land cover parcels (Figure 10). For example, in some areas we find that specific land cover categories occur consistently as neighbors. Often pastures and cropped land are found as neighbors. We do not expect to find cropped land within a densely urbanized area. These kinds of spatial patterns are important because they may permit the interpreter to overrule other visual clues in the identification of a parcel. The existence of patterns of occurrence between different categories can be exploited in the design of the classification system by tailoring category definitions to match their occurrence on the landscape. If cropland and pasture occur in individual parcels too small to be mapped, yet are adjacent to one another, the category 'cropland and pasture' permits mapping of these areas. If they do not occur as neighbors, however, the combined category may be of little use.

Site.   Objects or features that occur in characteristic geographic or topographic positions can frequently be identified by a knowledge of the significance of site as an element of image intepretation. For example, sewage treatment plants occupy low topographic positions, usually adjacent to water bodies. This kind of knowledge permits the interpreter to eliminate, or to confirm, tentative identifications based upon other evidence.

Association.   Some objects are so consistently associated with related objects that the identification of one indicates the presence of the other. A classic example from military photo interpretation is the confirmation in 1962 of the presence of Soviet missiles in Cuba (Goddard 1969). Initial interpretations did not identify the missiles themselves, but rather the construction patterns and support equipment that were known from prior experience to be associated with missile support units (Abel 1966; Hilsman 1967). This inference motivated the acquisition of photography that provided clear evidence of the missiles themselves. Babington-Smith (1974) described similar examples from World War II, including an account of the photo-interpretation search for German *Vergeltung* weapon sites in western Europe.

In an environmental sciences context, the principle of association is illustrated by the recognition of silty deposits at the mouth of a stream, which might lead an interpreter to search the drainage basin for the disturbed or eroded area that provides sediment for later deposition downstream.

For the interpretation of land cover, the use of association does not usually focus as much upon identification of parcels as it does upon the design of effective classification systems. When individual parcels are too small to be represented at the scale of the survey, and must be mapped within another category, it is sensible to design the classification so these small parcels are consistently grouped with the same category. This permits the map user to be able to predict the occurrence of impurities within the map categories. To be able to apply this strategy, the interpreter must be able to apply a knowledge of the association of separate categories within the mapped region.

Methods of Image Interpretation

An image interpretation method can be defined as a disciplined procedure that enables the interpreter to relate geographic patterns on the ground to their appearances on the image. Image interpretation methods can be divided into five categories (Campbell 1978).

Field observations, as an approach to image interpretation, are required when the image and its relationship to ground conditions are so imperfectly understood that the interpreter is forced to go to the field to make an identification. In effect, the analyst is unable to interpret the image from knowledge and experience at hand, and must go to the field to ascertain the relationship between the landscape and its appearance on the image. Field observations are, of course, a routine dimension to any interpretation as a check on accuracy, or a means of familiarization with a specific region. Here, their use *as a method of interpretation* emphasizes that when they are required for interpretation, field observations reflect a rather primitive understanding of the application of the remote sensing system for a specific subject.

Direct recognition is the application of an interpreter's experience, skill, and judgment to associate the image patterns with informational classes. The process is essentially a qualitative, subjective analysis of the image using the elements of image interpretation as visual and logical clues. In everyday experience direct recognition is applied in an intuitive manner; for image analysis, it must be a disciplined process, with very careful systematic examination of the image.

Interpretation by inference is the use of a visible distribution to map one that is not itself visible on the image. The visible distribution acts as a surrogate, or proxy (i.e., a substitute) for the mapped distribution. For example, soils are typically defined by verticle profiles that cannot be directly observed on remotely sensed imagery. But soil distributions are sometimes very closely related to patterns of landforms and vegetation that are recorded on the image. Thus, landforms and vegetation can form surrogates for the soil pattern; the interpreter infers the invisible soil distribution from those that are visible. Application of this strategy requires a complete knowledge of the link between the proxy and the mapped distribution. Attempts to apply imperfectly defined proxies produce inaccurate interpretations.

Probabilistic interpretations are efforts to narrow the range of possible interpretations by formally integrating non-image information into the classification process, often by means of quantitative classification algorithms. For example, knowledege of the crop calendar can restrict the likely choices for identifying crops of a specific region. If it is known that winter wheat is harvested in June, the choice of crops in interpretation of an August image can be restricted to eliminate wheat as a likely choice, and thereby avoid a potential classification error. Often such knowledge can be expressed as a statement of probability. Possibly certain classes might favor specific topographic sites, but occur over a range of sites, so a decision rule might express this knowledge as a .95 probability of finding the class on a well-drained site, but only a .05 probability of finding it on a poorly drained site. Several such statements systematically incorporated into the decision-making process can improve classification accuracy.

The final method of image intepretation is *deterministic interpretation,* the most rigorous and precise approach to image interpretation. Deterministic interpretations are based upon quantitatively expressed relationships that tie image characteristics to knowledge of ground conditions. In contrast with the other methods, most information is derived from the image itself. Photogrammetric analysis of stereo pairs for terrain information is a good example. A scene is imaged from two separate positions along a flight path and the photogrammetrist measures the apparent displacement. Based upon his knowledge of the geometry of the photographic system, a topographic model of the landscape can be reconstructed. The result is therefore the derivation of precise information about the landscape using only the image itself and a knowledge of its geometric relationship with the landscape. Relative to the other methods, very little non-image information is required.

Interpretations of land use and land cover can, and have, used a variety of image interpretation methods, including field observations, direct recognition, interpretation by inference, and probabilistic interpretation, depending upon the character of the study and the resources available. As a generalization, though, it must be recognized that a significant characteristic of land use/land cover interpretations is that land cover forms a proxy for the ultimate subject of the interpretation, land use. The interpreter can only map those features that are visible on the image, and must apply inference to portray the pattern of land use that lies behind the visible landscape. Dependence upon inference is widely used, and is accepted by those who make and those who use the resulting maps. Both users and interpreters must recognize the opportunities that exist for error and misunderstanding.

Interpretation Tasks

An interpretation may include a variety of tasks requiring examination of an image. The most basic tasks are *enumeration* (the listing and counting of discrete objects visible on an image) and *delineation* (outlining of boundaries between distinctive areas). Interpretations sometimes also require *mensuration,* the measurement of lengths, areas, or volumes from the image representation of an object or feature.

In the interpretation of individual objects, it is convenient to distinguish between several levels of knowledge. *Detection* refers simply to the determination of the presence or absence of an object or feature ("there is an object in the field"). *Recognition* represents a higher level of knowledge about the object (the object is determined to be a motor vehicle). And *identification* represents a level of knowledge detailed enough to assign the object to a specific class (the object is identified as a Ford pick-up truck). In an interpretation focused soley upon identification of objects, the interpreter can express his or her confidence in the identification by qualifying the results as 'possible,' or 'probable.' For interpretations of land use and land cover the focus is, of course, upon areal features, not objects, and the emphasis is usually not upon the identification, but upon the quality of the delineation. As a result, it is more difficult for the interpreter to convey his or her confidence in the results. Although there are few precedents for such a practice, it is conceivable that an interpreter could signify the presence of a particularly diffuse or indistinct boundary by a dashed or dotted line as a boundary symbol. The written report should, of course, explain the use of any symbols, especially those not commonly applied.

Collateral Information

Collateral, or ancillary, information refers to non-image information used to assist in the interpretation of an image. Actually, all image interpretation uses collateral information in the form of the implicit, often intuitive, knowledge, as well as formal training. In the more usual, narrower, meaning of the phrase, collateral information refers instead to the explicit, conscious effort to employ maps, books, statistics, and similar material to aid in the analysis of an image. Use of collateral information is permissable, and certainly desirable, provided two conditions are satisfied. First, the use of such information should be explicitly acknowledged in the written report; and, second, the information must not be focused upon a single portion of the image or map to the extent that it results in uneven detail or accuracy in the final map. For example, it would be inappropriate for an interpreter to focus upon acquiring detailed knowledge of tobacco farming in an area of mixed agriculture if he or she then produced highly detailed, accurate delineations of tobacco fields, but mapped other fields with lesser detail or accuracy.

Collateral information can consist of information from books, maps, statistical tables, field observations, or other sources. Written material may pertain to the specific geographic area under examination, or, if such material is unavailable, it may be appropriate to search for information pertaining to analogous areas — similar geographic regions (possibly quite distant from the area of interest) characterized by comparable ecology, soils, landforms, climate, or vegetation.

Image Interpretation Keys

Image interpretation keys are valuable aids for summarizing complex information recorded on film images. They are widely used in some aspects of manual image interpretation and remote sensing (Landis 1955; Heath 1956; Coiner and Morain 1971). Such keys serve either or both of two purposes: as a means of training inexperienced personnel in the interpretation of complex or unfamiliar topics, and as a reference aid for experienced interpreters to organize information and examples pertaining to specific topics.

An image interpretation key is simply reference material designed to permit rapid and accurate identification of objects or features represented on aerial images. A key usually consists of two parts: (a) a collection of annotated or captioned images or stereograms, and (b) a graphic or word description, possibly including sketches or diagrams. These materials are organized in a systematic manner that permits retrieval of desired images by (for example) date, season, region, or subject.

Keys of various forms have been used for many years in the biological sciences, especially botany and zoology. These disciplines rely upon complex taxonomic systems that are so extensive that even experts cannot master the entire body of knowledge. The key therefore is a means of organizing the essential characteristics of a topic in an orderly manner. It must be noted that scientific keys of all forms require a basic familiarity with the subject matter. A key, then, is not a substitute for experience and knowledge, but a means of systematically ordering information so that an informed user can learn quickly.

Keys were first applied to aerial images on a large scale during World War II, when it was necessary to train large numbers of inexperienced photo-interpreters in the identification of equipment of foreign manufacture and in the analysis of geographic and ecologic regions far removed from the interpreter's experience. The interpretation key then formed an effective way of organizing and presenting the expert knowledge of a few individuals. After the end of the war, interpretation keys were applied to many other subjects, including agriculture, forestry, soils, and landforms. Their use has been extended from aerial photography to other forms of remotely sensed imagery. Today interpretation keys may be used for instruction and training, but they have somewhat wider use as reference aids. Also, it is true that construction of a key tends to sharpen one's interpretation skills and encourages the interpreter to think more clearly about the interpretation process.

Keys designed solely for use by experts are referred to as technical keys. Nontechnical keys are those designed for use by those with a lower level of expertise. Often it is more useful to classify keys by format and organization. *Essay keys* consist of extensive written descriptions, usually with annotated images as illustrations. A *file key* is essentially a personal image file with notes; its completeness reflects the interests and knowledge of the compiler. Its content and organization suit the needs of the compiler, so it may not be organized in a manner suitable for use by others.

Materials and Equipment

The fundamental process of image interpretation can be completed without the use of elaborate or expensive equipment. In general, the interpreter should be able to work at a fairly large, well-lighted desk or work table with convenient access to electrical power. Sometimes it is useful to be able to control lighting with blackout shades or dimmer switches. Basic equipment and materials include a supply of translucent drafting film, some form of magnification, stereoscopes, tube magnifiers with measuring reticles, an engineer's scale, together with protractors, triangles, dividers, and other drafting equipment. A light table is important if transparencies are used. If roll film is employed, the light table must be equipped with adjustable brackets and an assortment of empty spools for take-up reels. Maps, reference books, and other supporting material should be available as required. More expensive items, such as binocular microscopes and binocular stereoscopes, are desirable but may not be essential.

If the interpretation is made from paper prints, special attention must be devoted to the prevention of folding, tearing, or rough use that will cause the prints to become damaged. Usually it is best to mark an overlay registered to the print rather than the print itself. Drafting tape must be selected specifically for its weak adhesive quality which will not tear the emulsion. The stronger adhesive used on many of the popular brands of paper tapes will damage paper prints.

If transparencies are used, special care must be given to handling and storage. The surface of the transparency must be protected by a transparent plastic sleeve or handled only with clean cotton gloves. Moisture and oils naturally present on unprotected skin may damage the emulsion, and dust and dirt will scratch the surface. Damage invisible to the naked eye may be a major problem under magnification.

The Image Overlay and Final Map

The interpretation process (Figure 11) begins with the assembling of imagery, collateral information, equipment, and materials required to conduct the interpretation. From a broader perspective, one could argue that the process actually begins earlier with the selection of image date, format, scale, and so on, as these qualities will greatly influence the character of the interpretation, even though the image interpreter may not

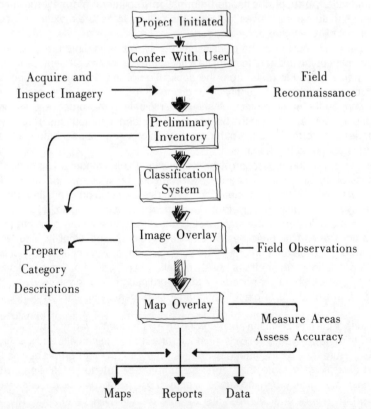

FIGURE 11 SEQUENCE FOR MANUAL INTERPRETATION. This diagram illustrates the sequence of some of the most important steps in manual interpretaton for land use mapping. Actual procedures will, of course, vary to meet requirements of specific projects, but most of the tasks represented here should be included.

always control these variables. Imagery should be inspected to note defects, gaps in coverage, and areas obscured by clouds. If there is no index to the coverage of separate frames or images, the interpreter should prepare his own index to permit convenient identification of specific images covering a given area. If there are gaps in the coverage, the interpreter should initiate the process of acquiring additional imagery to provide the best possible information for voids in the primary coverage.

The interpretation process consists of two related but distinct steps. The first consists of constructing an appropriate classification system. Although many different approaches to classification of land use can be applied, the system proposed by Anderson and colleagues (1976) provides one of the best and most widely applicable outlines for structuring the classification. Their system (Table 7) provides the more general categories, but the interpreter must become involved in the selection of more detailed categories at levels II and III. The definition and design of these detailed categories must be accomplished in co-ordination with those who will ultimately use final maps and data. If it is determined that the users require information or detail not feasible with the use of aerial imagery, then the project must turn to other imagery, or to collateral information, to supply the needed information. If collateral information is used, the interpreter should carefully investigate the alternatives to assure maximum accuracy and compatibility with the aerial imagery.

When a tentative outline of the classification is available, the interpreter should then carefully inspect the imagery to compile a list of the categories present; estimate the sizes of parcels likely to result from the application of the classification; and, in general, anticipate problems in identifying or delineating categories. If several interpreters will work on the same project, all should participate in this process to assure uniformity of perspective, and gain the benefit of independent contributions. A revised list of categories then forms the basis of further discussions with the user before a final classification is accepted for the project.

The second step is the application of the classification to the imagery. This consists of the process of marking the boundaries between categories as they occur on the imagery, following consistent guidelines. Each separate parcel is outlined, then identified with a symbol (usually one to three numerals) corresponding to taxa in the classification system. Important general principles are the consistent use of a standard minimum parcel size and the principles of consistency, clarity, and legibility. If several interpreters work on separate portions of the same project, special attention must be devoted to coordination of their efforts to assure uniform application of the guidelines. Individual interpreters share responsibility for co-ordination with those working on adjacent areas to be sure that parcel boundaries and identities match at edges of sheets.

The results of the interpretation are recorded on a translucent overlay that registers to the image; this *image overlay* records the boundaries between parcels, and identifying symbols (Figure 12). As the image overlay is prepared, the interpreter begins to compile the *classification table* that forms a summary of category definitions and identifying characteristics. When the image overlay is complete, it will be necessary to generate a *map overlay* that portrays the land cover/land use information on an accurate planimetric base.

An important principle, not immediately obvious, in preparing land cover maps from aerial imagery is that the map cannot stand by itself. A written report that summarizes the interpretation process and presents a clear definition and description of each mapping unit must accompany each map. Without the written report the

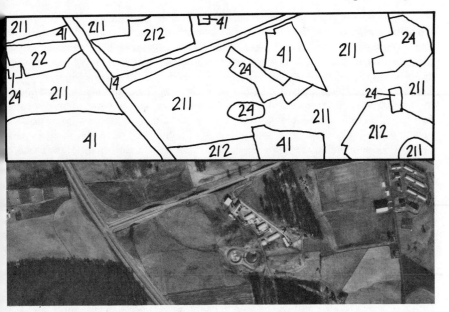

FIGURE 12   THE IMAGE OVERLAY. The overlay at the top is derived from the photograph shown below. The image overlay is a working document of the photointerpreter used to record parcel boundaries and identities. Symbols refer to Table 7.

reader has no means of understanding the classification system and how it has been applied to the imagery. The interpreter should probably devote as much time to recording and describing the map and its preparation as to preparing the map itself.

Preparation of the Image Overlay

The following paragraphs outline some of the essential considerations in preparing the image overlay (Figure 12) and in recording information for the accompanying report. Some of these guidelines follow from an obvious interest in visual and logical clarity; others are simply conventions established by long application of rather arbitrary decisions. For the novice, it is probably best to follow the guidelines as outlined below; later, as additional experience is acquired, variations can be devised as required. But always the guiding principles should be attainment of visual and logical clarity, and explicit description of the interpretation process.

1.   Identification of land cover parcels is based upon the elements of image interpretation as discussed previously. Sometimes interpretation may require identity of specific objects or facilities, but usually the primary task is one of *consistent delineation of uniform parcels* that match the classification system. The greater the uniformity of actual land use within areas represented on the map under a single symbol, the greater the usefulness of the map to the user (Webster and Beckett 1968). As outlined below (items 9 and 11), it is often necessary to violate this principle, especially at small scales.

2. The classification system must dovetail with categories accepted as useful b' the map user. Maps and reports that organize information in a manner inappropriate fo users' requirements are of little practical use; the image analyst has responsibility fo assuring that the final product is consistent with users' needs or for advising the use that the desired information cannot be reliably derived from aerial imagery.

Sometimes classifications are proposed that tailor categories in respect to ease o recognition on specific forms of imagery. For the kinds of user-oriented maps discussec here, such classification systems would not be suitable, because they organize infor mational content of the map around the image, rather than upon the users' require ments. (Remember that the users' requirements may be dictated by legal definitions, by policy of a governing body, or by requirements for compatibility with data collectec previously or by a neighboring jurisdiction. User specifications are not easily redefinec to suit desires of the image interpreter.)

As a result, the interpreter should not define categories solely from appearance or the imagery. For example, categories such as "suburban land" and "strip develop ment" are probably poor categories for most land use maps because they are basec largely upon ease of recognition on the image, rather than correspondence with the land use categories of interest to most planners and geographers. Thus, "suburbar land" should be redefined and mapped, as appropriate, into specific residential anc commercial land use categories accepted in the user's definitions. (Note that any category, if properly defined and tailored to users' needs, may have its merits; as a rule however, it is best not to devise unconventional categories for the convenience of the interpreter.) In brief, land use categories used in making the map should have clear meaning to those who will use the map.

The USGS classification system (Anderson *et al.* 1976) need not be perceived as the ultimate classification (for example, see Drake 1977), but for most purposes i' seems to form a good standard unless there is good reason to use another system Nunnally and Witmer (1970) have observed that one of the problems with land use classification prior to the proposal of the USGS system was the incompatibility of the many systems then in use. The USGS system seems to have taken a big step in solvinc that problem, as it forms a widely accepted framework for classifications at varied levels of detail.

3. Use of collateral material may be necessary if the interpreter is not intimately familiar with the region, or if unusual categories are encountered. Non-image collatera information might include topographic maps, existing land cover maps (at scales o' dates differing from the one in preparation), or tabulated economic statistics. Additiona imagery at large scale can be used to resolve uncertainties emerging from analysis o' small-scale, coarse resolution imagery.

4. The image overlay records, in manuscript form, the boundaries between lanc cover parcels. Each parcel is completely enclosed by a boundary, and is labeled with a symbol keyed to the category descriptions in the classification system. As a genera rule, the image overlay shows only those features that occupy *areas* at the scale of the final map. Usually, point or linear features are not mapped. Thus, a highway would no' normally be shown unless publication scale permits legible delineation of both sides o' the highway right of way. For example, at very small scale, even four lane expressways will be represented (if at all) as lines; at somewhat larger scale the cloverleaf inter changes are large enough to be mapped. A large scale land cover map might be able tc show the entire highway as an areal feature. This same rule should be applied to othe linear features, such as streams, railways, and power lines. Selected point or linea features may, of course, be useful as landmarks or locational references on the map, but they are not classified as areas.

5.   The principle of consistent composition in each category means that the map user can be confident that the map presents a consistent representation of variation present on the landscape. Each parcel will, by necessity, encompass areas of categories other than that named by the parcel label; these inclusions are permissible, but must be clearly described in the category descriptions, and must be consistent throughout the map. The issue of consistency is especially important when several interpreters work on the same project. Individual interpreters must coordinate their work with those working on neighboring areas, and supervisors must check to be sure that detail is uniform throughout the mapped area.

6.   The entire area devoted to a specific use is delineated on the overlay. Thus, the delineation of an airfield normally includes not only the runway, but also the hanger, passenger terminals, parking areas, access roads, and all general features inside the limits of the perimeter fence (the outline of the parcel encompasses areas occupied by all of these features not shown individually on the map). In a similar manner, the delineation of an interstate highway includes not only the two paved roadways, but also the median strip and the right of way.

7.   The issue of multiple use is discussed in detail by Anderson and colleagues (1976). In brief, the problem is caused by the practice of assigning parcels to single categories even though we know that there may in fact be several uses. A forested area may simultaneously serve as a source of timber, and as a recreational area for hunters and hikers. In general the interpreter must make a decision, apply it consistently throughout the image, and clearly document the procedure in the written report.

Numerous variations on this problem are encountered frequently. For example, employee parking areas within a large industrial area could logically be classified either as industrial land (applying the 'entire area' rule mentioned above) or as a transportation feature. The important considerations are usually that the decision fit user needs and be consistently applied and clearly documented.

8.   The interpreter must select an appropriate minimum size for the smallest parcels to be represented on the final map. The interpreter may be able to identify on the image parcels much too small to be legibly represented. Therefore, it is necessary to select a minimum size for the smallest parcels on the map. If several interpreters are working on the same project, all must apply the same minimum parcel size to assure that the variations in map detail reflect actual variations in parcel size on the ground.

In determining an appropriate minimum size for land cover parcels, it is important to remember that it is the minimum parcel size on the final map overlay and the legibility of the map (rather than the image overlay) that is of interest to the user. Because the map overlay may be presented at a different scale than the image overlay, the interpreter must extrapolate the minimum parcel size from the scale of the final map to the working scale of the image. Because it is difficult to perform this extrapolation mentally, it is sometimes useful to prepare a rough template of the approximate correct size to aid the interpreter in maintaining the correct level of spatial detail on the image overlay.

Anderson and colleagues (1976) recommended that parcels on the final map be no smaller than 0.1 in (2.54 mm) on a side (about 6.5 sq mmm). Loelkes (1977:17-18) presented values for minimum sizes of parcels to be represented on USGS land use maps. Parcels of urban land, water, and certain other specified categories are to have minimum sizes of 4 hectares(10 acres). At 1:100,000 such parcels would occupy about 4 sq mm on the map; at 1:250,000 they would require about 0.64 sq mm. For all other categories, the minimum size should be 16 hectares (40 acres). At 1:100,000 these parcels occupy about 16 sq mm, or about 2.6 sq mm at 1:250,000. These guidelines apply for rather compact parcels; Loelkes proposes additional guidelines governing minimum widths for long, narrow delineations. Although these sizes may be appropriate for the USGS land cover maps, they seem too small for routine use by the interpreters using equipment and procedures that may differ from those available for USGS interpreters. This same topic has been discussed by pedologists in relation to soil maps. Because the issue is essentially one of visual legibility, the informational content of the map should not influence our choice of minimum sizes of parcels, and the conclusions of soil surveyors may be of some interest.

Boulaine (1980) recommended that the smallest parcels on a published soil map should occupy areas no smaller than 25 sq mm if the parcel is square in shape, or 6.16 sq mm if circular. The smallest map distance separating two parallel lines should be at least 2 mm. These guidelines are in approximate agreement with those of Fridland (1972), who specified that the smallest delineations on published maps should be no smaller than 20 sq mm. The values that Kellogg and Orvedal (1969:125) recommended as minimum sizes for parcels on published soil maps translate to a map area of about 40 sq mm.

The minimum sizes suggested by soil surveyors for soil maps are larger than those suggested by Loelkes for USGS land use maps. Although most of the USGS maps have adequate legibility, many interpreters may prefer to use minimum sizes larger than those proposed by Loelkes. It should be obvious, however, that any single value proposed as a minimum size for mapped parcels should be interpreted as appropriate for the occasional presence of small parcels; if the interpreter encounters an extremely complex pattern of extremely small units, then an attempt to represent them all at the sizes mentioned above would produce an illegible map. In such a situation, some form of generalization is clearly required to produce a visually and logically clear map at publication scale.

9. Usually the label of each category identifies the *predominant category* present within each parcel. At small mapping scales, especially, there may be inclusions of other categories; the mapping effort should aspire to define categories that include relatively consistent mixtures, identities, and proportions of such inclusions, and to describe accurately their presence within each category.

10. Sometimes the identification or correct placement of boundaries can be a problem, especially if the interpreter can discern a wide transition zone between categories. Usually the interpreter can place the mapped boundary at the center of the transition zone, then describe the situation in the written report that accompanies the map. Sometimes wide transition zones occur consistently within the mapped area; if so, it may be appropriate to define a separate category: "416. Transitional Zone Between Evergreen and Deciduous Forest Land."

11. *Mosaics* of contrasting categories can present problems if the individual parcels are too small to be represented legibly at the scale of the final map. In these situations, it may be appropriate to create a category tailored to describing the situation: "215. Mosaic of Cropland and Pasture." The written category description then specifies the sizes and shapes of the parcels, and presents an estimate of the percentages of the areal extent of each member of the mosaic. These composite mapping units are often a necessary departure from the principle of uniform mapping unit composition outlined above, but there is ample precedent for their use (Christian 1959; Robinove 1981).

Mapping Land Use Change

Land use patterns change over time in response to economic, social, and environmental forces. The practical significance of such changes is obvious. For planners and administrators they reveal areas that require the greatest attention if communities are to develop in an harmonious and orderly manner. From a conceptual perspective, study of land use changes permits identification of long-term trends in time and space, and the formation of policy in anticipation of the problems that accompany changes in land use (Estes and Senger 1972; Anderson 1977; Estes *et al.* 1982).

A map can show only a single temporal image of the many that form the evolving pattern of land use in a region. As a result, any land use map is inaccurate almost from the time it is prepared. Users who are familiar with the mapped region will accumulate an informal knowledge of changes that have occurred after the map was prepared — a mental map of the changes. For systematic study of changes, however, it is necessary to prepare maps that formally document changes in land use between two specific dates. Aerial imagery provides the unique capability to reconstruct previous land use patterns using archived images, even though no map was prepared at the time. In some instances, this capability acquires great significance, as, for example, in the use of archival aerial photography to document land use changes at abandoned hazardous waste sites (Schweitzer 1982).

In theory, preparation of change maps is very simple. However, a number of practical problems are encountered in practice. Preparation of a change map requires comparison of two separate land use maps prepared from imagery acquired at two dates. Areas which experience changes in land use are noted (usually by superimposition of the two maps), then recorded on a third map. This third map shows only the changes, which can then be tabulated by area and by category to reveal the extent and location of land use changes. If imagery for several dates is available, a series of change maps can record the evolution of land use patterns over time and possibly reveal long term patterns of change, rates of change in specific areas, and intermediate steps in the development of land use patterns.

Although this procedure is essentially straightforward, a number of practical problems must be anticipated. First, the two maps must share a common base before they can be registered to one another. Even if both maps use a common base, the interpreter must work carefully, as minor differences in placement of boundaries or different levels of detail can create differences that are not evidence of land use change, but merely artifacts of the interpretation process. For these reasons, it is important that the same interpreters work on all phases of a change map, or that interpreters be supervised by a single individual with overall knowledge of the project. Preparation of change maps also requires that both maps be prepared using a single classification system applied consistently at a given level of detail.

These considerations lead to the general observation that preparation of change maps requires continuity in technique and in personnel, and close coordination of the mapping process. Source maps must be compatible in respect to classification, spatial detail, and map base. As a result, one should be skeptical of maps prepared by different individuals or organizations used as the basis for maps of change.

The image interpreter may find that preparation of a change map requires examination of imagery at varying scales, resolutions, and qualities. Each interpreter must assess the imagery in relation to map scale and classification detail to determine the level of detail most suitable for the change map. As a general guideline, it seems sensible to recommend that all change maps be prepared at a level of detail consistent with that obtained from interpretation of the lowest-quality, coarsest-resolution imagery that will be used in the project. Otherwise, the interpreter will be faced with the problem of comparing two maps that differ greatly in detail and accuracy.

Field Observations

Even the most thorough, accurate interpretations require use of field observations as confirmation of manuscript maps, and as a means of resolving uncertainties in the interpretation process. Ideally, field observations should be acquired on at least three occasions during preparation of the land use map: (a) During preparation of the classification system, before the interpretation of imagery begins, as a means of familiarizing interpreters with the region and its major land uses; (b) as the image overlay nears completion, to verify uncertain interpretations and to confirm consistency of the interpretation; and (c) when the preliminary draft of the map overlay is complete, to detect and resolve any final problems before the final copy of the map is prepared. These three excursions to the field serve different purposes, so they may vary in respect to duration, route, and intensity.

Although details of field excursions will depend greatly upon individual preferences and local circumstances, the following observations seem generally applicable:

(1) Imagery should be taken to the field, together with manuscript overlays and supporting notes and maps. Provision must be made for annotating images (on overlays) or correcting manuscript overlays in the field. This usually means that maps and overlays should be temporarily clipped to a hardboard or cardboard surface small enough to carry in the field and to use in a vehicle, but large enough to present a sizeable portion of the map for navigation and annotation.

(2) The route should be planned carefully to select an efficient itinerary that covers all essential areas. If timing is critical, it is important to allow time for unexpected delays. It may be wise to assign priorities to specific areas, so that the most important areas can be visited first.

(3) Notes, photographs, and sketches should be made in a systematic manner that ensures their usefulness later in the lab. If machine copies of maps and aerial photographs are made beforehand, they can be used as a medium for making notes, and recording locations of photographs and stops. These images will probably not record all detail visible on the original, but they are usually suitable for making notes without damaging the original.

(4) If several interpreters are participating, it is usually best to work in teams, with division of labor such that each has responsibility for specific tasks. If interpreters have been assigned specific geographic areas, then each is responsible for planning the itinerary and assumes direction of excursion for their area.

(5) If the project has not already initiated contact with local planning, soil conservation, and agricultural extension personnel, these

organizations should be involved in providing advice and information. They may be especially helpful in planning field excursions.

(6) At the completion of the trip, a systematic effort must be made to
organize all material and information, and to be sure that field
notes are clearly transcribed into a more formal format while their
meaning is still clear to all involved. Notes and annotated maps
must be identified in respect to date, and the person who made
the observations.

Checking and Editing

Errors are inevitable. As a result, checking and editing are essential steps in the
preparation of a map, equal in significance to the more immediately obvious steps of
acquiring imagery and preparation of the image overlay. (We should not accept the
false distinction between those who make errors and those who do not, but rather
accept the more valid distinction between those who inspect for errors and those who
never look!)

The search for errors is continuous through the preparation of the map, but tends to
focus at specific stages. Each parcel on the image overlay should be compared to the
original image to confirm its identity, correct boundary placement, and adherence to the
minimum parcel size. On the final draft of the map overlay, a check is made to confirm
the presence and legibility of all boundary segments, labeling of parcels with their
correct symbols, and registration of land use detail to detail on the base map. For large
projects with several separate sheets, a specific check must be made to ascertain that
boundaries match at the edges of sheets, and that parcel identities match across sheet
edges.

There is little benefit to be gained from an attempt to list all of the errors that can
occur in the preparation of a land use map from aerial imagery. But it may be useful to
propose two principles that can be applied to detect and reduce errors of all kinds. First,
checks for errors should be made throughout the preparation of the map; the earlier
errors are detected, the easier they are to correct. However, if the check for errors is not
focused at specific stages in the production of the map, the search for errors becomes
so diffuse that it loses meaning. Therefore, specific steps in the production process
(perhaps at the completion of the image overlay and completion of the map overlay)
should be designated as opportunities for checks of the work completed thus far — a
hurdle that must be passed before the next step begins. Second, errors are easier to
control if specific individuals or groups assume responsibility for specific portions of the
project. Checking and editing should then be clearly separated from the preparation
process by designated individuals to check the work of others, or having individuals
check their own work at a time and place different from the time and place of original
preparation. These steps assist in promoting a critical attitude during the process of
checking work for errors.

The Map Overlay

The map overlay is formed by plotting boundaries from the image overlay onto an
accurate planimetric base (Figure 13). Parcel boundaries on the image overlay include

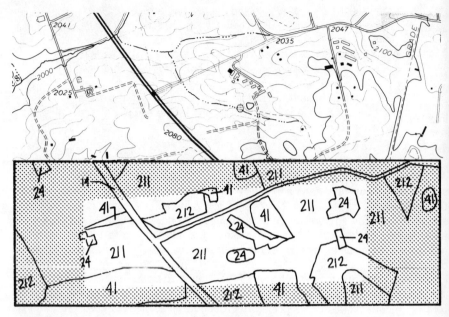

FIGURE 13   THE MAP OVERLAY. This overlay, prepared from the overlay in Figure 12, registers to the planimetrically accurate map shown above. The land use boundaries have been changed in scale and geometry so they now have a form suitable for measurement of areas or distances. The shaded pattern outlines the approximate area shown in Figure 12, which is now represented at a smaller size because of the change in scale. Symbols refer to Table 7.

positional errors inherent to all remotely sensed images. The image overlay, a preliminary document, cannot be used as the basis for accurate measurements of distance or area. The map overlay registers to an accurate map, so that corresponding detail (when present) matches exactly; it forms the basis for the final land use map. Even in instances when geometric errors have been minimized by preprocessing of the remotely sensed data, it is often desirable to plot land use boundaries on a map base compatible with other maps frequently used.

Changes in image scale and geometry can be made using a number of methods, including use of reflecting projectors, or instruments comparable to the Sketchmaster or the Bausch and Lomb Zoom Transfer Scope. Often preparation of the map overlay requires not only changes in map scale, but also changes in map geometry necessitated by geometric properties of the remotely sensed images. Devices such as the Zoom Transfer Scope are useful for this purpose because of their ability to permit convenient changes in image geometry, a process that is difficult with the usual reflecting projectors.

Regardless of the procedure to be used, the process of matching the two images is easiest when there is ample detail common to both image and map. As a result, the image overlay should be prepared to show, for example, locations of drainage, topographic, and transportation features visible both on the map and on the remotely sensed image, even if such features do not always correspond to land use boundaries shown on the final map. Such features provide the common detail necessary to permit convenient and accurate registration of the two maps.

The map overlay shows land use boundaries and symbols, plus other features (such as major highways, rivers, and place names) that may be useful to the reader in orientation and interpretation of the land use map. The completed map should include a bar scale, legend, title, coverage diagram, and other information required for accurate interpretation.

The Report

Maps seldom stand alone, to be accepted by readers at face value, without supporting information. The notion of "supporting information," in this context, should be interpreted broadly, to include not only formally presented written material that accompanies the map, but also the wider realm of knowledge that the reader uses to examine and evaluate a map. Some maps include written explanations of mapping technique and mapping unit characteristics, either as text on the map itself or in documents that accompany the map. This kind of information forms explicit information formally presented to the map reader.

Although many maps are without supporting information in such explicit form, the map reader often has the benefit of implicit, informal knowledge derived from experience with similar maps. For example, the reader of a USGS topographic quadrangle is presented with very little explicit information concerning mapping technique if the map itself is considered as the only source of information. However, the reader usually has access to substantial implicit information, acquired in the examination of other similar maps and through knowledge of cartographic conventions. As a result, the reader can employ in a wide range of knowledge not obvious from inspection of the map itself — for example, symbolization, accuracy, cartographic conventions, and degree of generalization.

Land use maps, despite many superficial similarities to one another, are characterized by notable diversity in respect to purpose, categorization, detail, accuracy, and symbolization. Therefore, the reader must depend largely on explicit information, formally presented in written documents that accompany the map. As a result, careful preparation of supporting documentation assumes an importance equal in significance to preparation of the map itself.

Specifically, the reader requires knowledge regarding: (a) the regional setting, (b) methods and materials used to prepare the map, (c) definitions of mapping units, and (d) the summarized results. Emphasis devoted to each of these topics may vary in relation to requirements of the organization that will use the study, purposes of the study, and the experience of the intended users. For example, the description of the regional setting may be abbreviated if it is known that the users of the map are already familiar with the area to be mapped. As a general rule, however, each item should be discussed, to provide a complete document that can stand by itself.

Regional Setting

The regional description outlines the geographic setting of the mapped region with emphasis on factors most likely to influence the development of land use patterns (Figure 14). Unless the area is unusually large and diverse, a brief narrative of a few

REGIONAL SETTING:

Roanoke is a major transportation center
for southwestern Virginia, having railroads,
major interstate highways, limited access highways, and an airport. Roanoke County falls in
the physiographic Blue Ridge Region of the
Eastern Farming and Forest belt. The Blue Ridge
Region extends from eastern North Carolina
through Virginia and Maryland and as far south
as South Carolina and Georgia containing
approximately 18,900 square miles. Typically,
two-thirds of the area is forested (20% in
National Parks and Forests) with 10% used in

FIGURE 14   WRITTEN REPORT: REGIONAL SETTING

paragraphs should be sufficient. The reader who requires a detailed description should
be directed to longer and more elaborate documents that focus upon specific aspects of
the region.

The purpose of this report section is simply to set the stage for subsequent
description and analysis of the land use patterns. These patterns can of course be best
understood in relation to the physical and economic context of the region. The physical
setting is described in brief outlines of climate, topography, soils, drainage, and natural
vegetation. The economic setting is described in terms of the key elements of the
industrial, commercial, and agricultural life of the region. In most instances this section
should also include a brief description of the regional transportation system, with
emphasis of links to other regions. The brevity of the regional description precludes
completeness; it should, however, sketch the main features of the regional economic
pattern, with emphasis upon the interplay of physical and cultural elements that determine the broad features of the regional land use patterns.

Methods and Materials

Also very brief, this report section consists ideally of only a few concise paragraphs
that describe the imagery and interpretation techniques used in preparing the map and
supporting data and documents (Figure 15). Imagery scale, resolution, date, quality,
format, coverage and source should be described. If it has been necessary to use

SOURCE MATERIAL:

The imagery used for this land-use/land cover overlay includes 9 inch color infrared transparencies (EDC scene ID: 530012152348, frames 2351-2353) and 18 inch paper prints produced from these transparencies. This imagery was obtained at a negative scale of 1:129,000 by NASA on 4 June 1973. The overlay registers to the northern half of the Roanoke 7.5 minute USGS quadrangle.

FIGURE 15   WRITTEN REPORT: SOURCE MATERIAL

several missions to complete coverage of the study area (due perhaps to gaps, clouds, or partial coverage by the primary imagery), the coverage diagram should depict respective coverage of each form of imagery (Figure 5). The character and sources of collateral information are also described, as are the chharacter and timing of field observatins. The interpreter also provides an account of the interpretation procedure, with mention of any special equipment used.

Mapping Unit Descriptions

Descriptions of mapping units form the most important part of the written report; they describe each category used in making the map, *as they have been defined for this specific report,* and as they have been applied to this specific image (Baker *et al.* 1979). Information should be presented concisely, clearly, and in sufficient detail to be of use to the reader in interpreting the map. Each category is described by specifying four separate elements (Table 9).

First, the name and symbol are presented exactly as they are used in the map legend and on the map itself. Every category used on the map appears in the written report.

Second, category definitions give precise, clear meanings. The reader may understand the general, conceptual definition of 'urban land,' for example, but cannot be expected to know the specific operational interpretation applied using specific imagery of a specific geographic region. Often the interpreter may be required to make very subtle or arbitrary distinctions in applying the classification to a specific image, and these distinctions may vary at differing scales and resolutions, and with differing forms of aerial imagery. Without the benefit of the information presented by the interpreter, the map reader has no means of reconstructing the operational meaning of the categories on the map. Descriptions should usually be concise; if elaborate descriptions are required, they probably should be presented in an appendix.

Third, a 'ground features' section presents an inventory of the primary objects and features that occur within each mapping unit. This section serves several purposes. It permits the reader to acquire a very precise understanding of the way that the interpreter has applied the classification to the image; in effect, this section reveals the interpreter's operational definition of each category. It also maintains the interpreter's discipline in defining mapping units; the interpreter who has difficulty in preparing concise inventories for each category discovers, in effect, that the categories have not been carefully defined or consistently applied to the image — lapses better discovered earlier rather than later.

TABLE 9   SAMPLE DESCRIPTIONS OF LAND-COVER CATEGORIES DEFINED FROM REMOTE SENSING IMAGERY

Symbol	Name	Definition	Ground Features	Image Appearance
111	single-unit residential	land occupied primarily by detached dwellings and associated structures	individual homes, lawns, streets, trees	tree crowns may be dominant features (medium dark tone, coarse texture); regular street pattern; driveways, sidewalks, lawns visible; rooftops visible
112	multiple-unit residential	dwelling units designed for occupation by several families	apartment buildings grassed areas, parking areas	large buildings usually rectangular, arranged in clusters; rooftops, parking lots visible; trees usually absent or sparse
211	cropland	land used for harvested cash crops	plowed farmland, land planted in crops, fencelines, hedgerows, farm roads	fields frequently have straight, or even sides; fine, even texture; photo tone usually light (very dark for plowed land); contour plowing, strip cropping frequently visible; field size usually small or moderate
212	pasture	land used primarily for grazing live-stock or for hay	open grassland, occasional isolated shrubs, trees, fence-lines, hedgerows, farm roads	fields frequently large irregular shape, indistinct boundaries; medium photo tones; texture is mottled in appearance

The most important function of the ground features section is to present an accounting, for each category, of the presence of foreign inclusions. For example, the limitations of mapping scale may require that unmapped parcels of forest be included within areas designated on the map as cropland. If so, the mapping unit description for cropland should specify the presence, identity, proportion, and (if possible) the pattern of occurrence of the unmapped inclusions. ("Includes small isolated patches of deciduous forest too small to be mapped up to a total of about 15 percent of the area mapped as cropland. Size and frequency of these areas decreases toward the southern edge of the mapped region.")

Fourth, the image appearance section describes each category *as it appears on the image,* using the "elements of image interpretation" as a framework for description. Table 10 lists a suggested vocabularly, with examples, as a means of describing the image appearance of land use categories, as they appear on black and white aerial photography. Variations can be devised to suit other forms of imagery, with a range of scales and resolutions.

The 'image appearance' section does not attempt to describe the ground appearance of the category, but the appearance of the category as it is represented on the imagery used for the study. In brief, it does not form a general, universally applicable description, but merely an account of the facts that apply to the specific interpretation at hand.

TABLE 10   VERBAL DESCRIPTIONS OF LAND USE CATEGORIES AS DEFINED FROM AERIAL IMAGERY

Often interpreters encounter difficulties in preparing written descriptions of the results of manual interpretations of aerial images. Here a variety of qualitative descriptive terms are listed as suggestions for your interpretations. These terms are in a sense imprecise, and must apply only to specific images, but they do offer a means of specifying image characteristics of land cover categories. Develop your own modifications as you gain experience.

Element of Image Interpretation	Some Suggested Qualitative Descriptors
Size:	"small," "medium, "large," *also:* is size: "uniform" or "varied"?
Shape:	"compact," "regular," "elongate," "square," "irregular," "rectangular"
Tone:	"light," "medium," "dark" "very light," "very dark"
Texture:	"coarse," "medium," "fine," *also:* "even," "uneven" "mottled," "uniform"
Association:	State if there exists a consistent spatial association with other categories. What is the character of the boundaries with neighboring categories?
Shadow:	Can you determine if shadow contributes to the appearance of a category? Consider not only objects, but *areas* as well (See Figures 7 and 8)
Site:	In some instances, topographic position (site) may be an important means of describing the distinctive characteristics of features or categories.
Pattern:	Specify if objects within a specific category are arranged in a distinctive manner. An obvious example: *an orchard.*

---

2   Agricultural Land

21   Cropland and Pasture - Parcels of varying
        shapes, sizes, tones, and arrangement which
        generally abut roads. Contain a mosaic of
        areas for the cultivation of agricultural
        products or livestock.

21.1   Cropland

            Areas with straight edges and rounded
            curves are aligned along roads and
            often associated in groups. Farm
            homes and buildings located within
            some clusters. Differences in fields
            recognizable by texture, tone, and
            color variations but particular crops
            not identifiable. Colors range from
            burnt orange to grey. Some parcels
            have same pink points recognizable as
            individual trees.

21.2   Pasture and Meadow or Pasture/Forest
            Regrowth

            Larger parcels often bordering crop-
            land. Mottled appearance. Tones
            vary from grey to grey/pink mixture.
            Tree clusters identifiable but too
            small to map.

FIGURE 16   WRITTEN REPORT: CATEGORY DESCRIPTIONS

Mapping unit descriptions can be presented in either of two formats. A brief narrative section, such as that used by Baker and colleagues (1979) presents each mapping unit description in a few concise sentences, organized to present all of the information outlined above (Figure 16). Or, it may be appropriate to present the same information in a table (the classification table) organized as illustrated in Table 9. The classification table serves two complementary purposes. First, it provides explicit information for the map reader regarding category definition and composition. Second, compilation of the classification table forms a means for the interpreter to evaluate the logic of the interpretation process. If an interpreter attempts to map ill-defined categories, or categories that cannot be clearly separated on the basis of image appearance, the problems quickly become evident in the preparation of the mapping unit descriptions.

Summary

The report concludes by summarizing the results of the inventory. Here the interpreter can describe problems encountered during the preparation of the study. If an evaluation of the map's accuracy has been conducted, results are reported here.

For most studies, however, the main portion of this section is devoted to summarizing the area occupied by each of the categories on the map. Because our interest here is essentially with inventories of existing land use, this summary forms a description of land use patterns as observed at the time of the imagery. Usually an evaluation or

TABULATION OF MAJOR LAND USE CATEGORIES

CATEGORY	LAND USE	AREA (IN ACRES)
41	Forest	5159.2
151	Undeveloped Urban	3564.1
111	Single Family Residential	3345.2
21	Cropland and Pasture	2108.2
22	Orchards, Groves, etc...	1154.7
146	Airport	707.6
121	Commercial and Services	690.3
76	Land in Transition	511.3
1732	Golf Courses	378.4
142	Four Lane, Limited Access Highway	220.0
212	Pasture	179.9
761	Road Cuts	148.6
163	Mixed Urban & Residential	137.5
53	Reservoirs	137.0
122	Primary & Secondary Ed.	135.9
171	Cemetaries	97.2
211	Cropland	80.8
112	Multiple Unit Residential	42.0
161	Mixed Commercial/Residential	55.9
762	Barren Areas	31.2
213	Pasture (overgrown)	48.3
23	Confined Feeding	29.2
145	Water/Sewer Treatment	26.1
54	Small Ponds	22.2
123	Large Shopping Centers	30.7
173	Recreational Land	14.5
113	Trailer Parks	10.0
	Total	19,065.1

FIGURE 17   WRITTEN REPORT: TABULATION OF AREAS.
Here, the categories are ranked in order of total area on the map.

interpretation of the appropriateness of the observed patterns is not appropriate in this context, although the results of the inventory may form the starting point for a separate study that does assess the relationship of existing patterns to ideal patterns.

A brief narrative may be appropriate, but the heart of the summary is a tabulation of the areas occupied by each land use category (Figure 17). The area of each parcel on the final map is measured using a planimeter, electronic digitizer, or other means. Accurate measurement of area depends on careful preparation of the map overlay; measurements made from the image overlay will be in error. Areas of separate parcels are summarized by category to yield a single total for each category. The final tabulation of areas shows each mapping unit by name and symbol, with its total area in the mapped region (reported in acres, square miles, or hectares as appropriate) and percentage of the total mapped area. Detailed categories are collapsed into broader categories, so the listing reports all possible levels of detail. If the mapped area has been subdivided into political or census units, areas of each category are also reported by subdivision.

Regardless of the method used to measure areas, the total for all categories inevitably differs from the known correct total for the mapped region. The interpreter is then faced with the problem of allocating the known error among parcels, and therefore among the several categories. There seems to be very little study of this problem, but one of the most practical procedures (if not the most accurate) is to allocate total error among parcels in proportion to perimeter length, on the assumption that measurement error increases with increased length of the line measured. (This procedure can only be a rough approximation because it seems clear that errors are also related to parcel size, shape, and complexity.) This procedure requires tabulation of perimeters until the total error is allocated. If a digitizer has been used, it may be a simple matter to store perimeter lengths for use in a computer program that finds and reports adjusted areas. If manual methods have been used, the interpreter will probably decide that practicality dictates use of parcel areas rather than perimeters as a basis for error allocation.

6

Machine Processing of Remotely Sensed Data

Although machine processing of remotely sensed data has been practiced for many years, it has found widespread use only recently. The launch of Landsat 1 in 1972 led to widespread interest in machine processing. Prior to Landsat 1, digital data were not widely available for the general user, although most of the major remote sensing centers were equipped to generate and analyse digital remotely sensed data. Among the wider community of remote sensing practitioners, the equipment and expertise required to routinely conduct automated interpretations were generally not available, partly due to the absence of digital remote sensing data.

As Landsat 1 began to generate a stream of imagery and data, it soon became evident that manual interpretation procedures would not be sufficient for full exploitation of information in Landsat data. With routine availability of digital data at modest cost to the user came increased interest among businesses, governmental agencies, and universities in developing capabilities for exploiting these data. Improved quality and increased availability of mini-computers, associated peripheral equipment, and computer software has also increased interest in digital processing among a broad cross-section of those who use remotely sensed data.

Today, the capabilities for digital analysis of remotely sensed data are dispersed widely in the remote sensing community. As a result, all who have an interest in remote sensing require at least a general understanding of the topic. Machine processing appears to be of special interest for those with an interest in land use mapping, because the most interesting developments in machine classification have focused on land use and land cover mapping.

Digital analysis encompasses a wide range of operations in which remotely sensed data are subjected to analysis or manipulation by algorithms that treat data and information in an abstract, formal manner. Despite the abstract quality of the analysis, and its apparent remoteness from direct human intervention, it is wrong to assume that an objective is to remove human judgment from the process. The analyst must make decisions throughout the process of machine interpretation, including choices regarding the kinds of operations performed, and specific options selected in implementing given operations. These decisions require skills and perspectives equivalent to those required for manual interpretation; there are many individuals who can implement the mechanical steps required to produce a machine-generated map, without the supporting knowledge and experience necessary to prepare a useful map. As a result, machine-generated maps are subject to the same kind of scrutiny we might apply to any map. That the map was generated by computer does not release it from the same standard we should apply to manually produced maps.

Advantages and Disadvantages of Machine Processing

Advantages of machine classification relative to manual interpretation are often seen in *cost effectiveness* for interpretation of large areas that must be repeated on a routine basis. For example, identifications of crops within large agricultural regions might be required several times throughout the growing season to record the kinds of crops planted, their relative areas, and growth states. This is the kind of task best suited for machine interpretation. Machine interpretations are most effective for single, uniform, images (such as Landsat imagery), or other data with standard formats. Machine interpretations can be awkward (at best) when it is necessary to use several forms of data with diverse characteristics. Machine analyses *yield consistent results:* given the same data, the same series of operations will yield identical output. Thus, the interpretation of an image can be exactly duplicated, or the same operation can be applied exactly to several images.

Digital classification can simultaneously interpret data in several spectral bands, and thereby conduct analyses impossible or impractical by manual methods. Often it is possible to construct composite data from several channels and dates, or to use ancillary data in a form compatible with that of the digital data. Likewise, abstract interpretation algorithms using statistical/quantitative transformations and decision rules are possible with machine interpretations. Once equipment and data are on hand and in operation, speed of interpretation can be an important advantage over manual methods. The ability to conduct analyses rapidly may permit the operator to explore alternative interpretations, and thereby acquire accuracy, flexibility, and a knowledge of the scene not possible with alternative approaches. Finally, output from machine classifications may be comptible with other digital data from other scenes or other sources, facilitating comparisons, compatibility with geographic data bases, and mapping of change.

Machine classifications have disadvantages and limitations. Machine classifications are expensive for small areas, or for analyses that must be conducted once or only infrequently. Start-up costs may be high, and lead time may be lengthy. Expensive equipment must be acquired, and personnel must be hired or retrained to service, maintain, and operate equipment.

There seems to be much more attention focused on the accuracy of machine classification than upon accuracy of similar manual interpretation. However, in many machine interpretations, quality and accuracy may be difficult to assess, so the user may be presented with a product of unknown quality. Moreover, machine analyses usually require standardized imagery and data. Often the most time-consuming step in conducting a machine classification is that of acquiring and then converting data into a format appropriate for the system to be used. In some instances this process may require more time and effort than the actual generation of the classification itself.

Despite the wide range of options available in most image processing systems, they can be said to be inflexible in some respects. For example, the system may be tailored to perform certain operations efficiently, but even modest departures from the original purpose may require substantial changes in system programs. Also, data may be expensive, unavailable in the desired format or date, or available only after long delay. Preprocessing may be required to correct errors present in the data. In a sense, the ability to conduct preprocessing is an advantage; seldom is preprocessing conducted for manually interpreted imagery. However, machine analysis may be more sensitive to small geometric and radiometric errors, and preprocessing may itself introduce undesirable characteristics into the data.

The Image Analysis Process

Machine analysis can best be understood in the broader context of overall data acquisition and analysis. For most of the data considered here, sensors react to energy emitted by the sun in the visible and near infrared portions of the spectrum. This energy passes through the earth's atmosphere, is reflected from the earth's surface, then passes back through the atmosphere before it reaches the sensor. Thus, the visible and near infrared energy has twice been subjected to effects of scattering and absorption in the atmosphere before it reaches the sensor. For images depicting very large areas, such as Landsat scenes, these effects can vary within a single image.

Energy reflected from the various features on the earth's surface can be understood only in the interaction between a specific landscape and the spatial, radiometric, and spectral resolution of a specific sensor system. These qualities determine the levels of detail or variation in a specific scene; the variability present depends, of course, upon the specific landscape to be imaged. Key variables include the number, sizes, and shapes of parcels present, and contrasts in brightness. Contrast in the landscape depends upon the identities of the parcels (forest, water, cropland, urban) and the time of year, to mention two of the most obvious factors.

Artifacts in remotely sensed data may be caused by spatial or temporal variations in solar angle, solar elevation, and local topographic slope, aspect, and relief. At the high altitudes and coarse resolutions of satellite-borne sensors, these elements assume a significance not evident in many other applications of remote sensing. Thus the character of the energy reaching the sensor is controlled by many diverse and interrelated variables, many unknown in detail to practitioners of remote sensing.

A given sensor may produce an image (in the sense of a photograph-like representation of the earth's surface), or data (an array of values), or both (Figure 18). Usually it is possible to change from one form to the other, if we are willing to accept some loss in quality. Once data are in hand, the image analyst conducts an interpretation using either manual or automated procedures. Usually the initial choice of data (either numerical or pictorial) is an implicit choice of method, so acquisition of numerical data assumes subsequent quantitative processing, but there is no reason that the image analyst could not use elements of both methods if necessary. We cannot judge that either approach is by its nature superior to the other. Both have characteristics suited for specific kinds of problems, and the best choice depends upon the resources available, the character of the problem at hand, training of personnel, and similar factors.

Digital Image Analysis

The following sections focus on the main features of numerical image analysis. Procedures for automated interpretations vary so greatly, depending upon the specific problems at hand and accepted procedures and programs established at specific facilities, that no concise explanation can cover all options available for an analyst.

```
18  15  20  18  15  16  21  21
17  16  18  18  16  19  21  19
15  18  16  17  16  20  19  17
14  19  19  19  16  18  16  18
17  18  18  18  15  17  18  18
21  18  17  17  14  16  19  20
22  19  17  17  14  16  18  21
20  18  17  17  16  18  20  21
21  17  19  19  19  20  21  19
19  15  18  20  21  22  20  15
14  13  15  19  22  22  16  10
12  11  15  20  21  17  12   9
14  14  18  21  17  10   8  12
16  17  19  19  14   8  10  16
18  17  15  15  12   9  14  18
15  11  13  14  12  12  17  20
11  10  15  17  12  14  19  20
12  16  20  19  13  14  20  20
14  20  21  17  13  16  17  16
18  19  20  17  15  17  17  19
18  16  19  20  17  18  23  26
15  18  20  19  18  23  25  24
16  16  16  16  16  20  24  23
17  12  12  14  13  16  21  19
14  13  14  13  14  19  17  14
13  13  14  12  13  16  14  11
13  13  13  10  13  13  12  11
13  12  12  10  13  13  13  15
15  13  11  12  13  14  14  17
18  13  12  15  17  15  16  17
```

FIGURE 18   PICTORIAL AND DIGITAL IMAGES. The upper left image is a portion of an aerial photograph (June 1980) of a warehouse complex near Roanoke, VA. The values below it show a small portion of the same area as recorded (November 1980) by the LANDSAT multispectral scanner (band 5). Each digital value represents the brightness of a ground area said to be approximately 1.1 acres (0.44 ha) in size. Low values indicate dark surfaces; high values indicate bright surfaces. The image at the upper right shows a display of these digital values with brightnesses scaled to represent the magnitude of the brightness recorded by the scanner. The blocky appearance is the result of the relatively coarse spatial resolution of the imaging system.

Carter and colleagues (1977) surveyed some of the options available on many of the larger image processing systems. Their list is now out of date, but still provides a useful summary of variations among systems. The following paragraphs therefore form only a rather rough outline of an idealized sequence for automated image interpretation (Figure 19). In actual practice details will vary greatly (Swain and Davis 1978; Moik 1980). Note that the sequence illustrated in Figure 19 does not replace Figure 11, but simply describes an alternative sequence for generating maps and data. The analyst should still follow the broader steps in Figure 11 that call for consultation with the user, field reconnaissance, and related steps.

Preprocessing

Preprocessing operations prepare data for subsequent analysis, usually by correcting, or compensating for systematic errors. Three classes of preprocessing operations can be defined. The first are those that simply display or summarize data as a means of inspecting quality and detecting the presence of errors. Frequency histograms, scattergrams, or statistical summaries permit the operator to assess image quality and determine subsequent preprocessing steps (if any) that may be necessary.

A second group of preprocessing operations are those that compensate for *radiometric errors,* errors in measurements of brightness. These errors result from defects in sensor calibration or operation, from atmospheric absorption and scattering, variations in scan angle, illumination of the scene, and from system 'noise.' Typical preprocessing to correct for atmospheric degradation include simple adjustments based upon reflectances of objects of known brightness. More complex algorithms

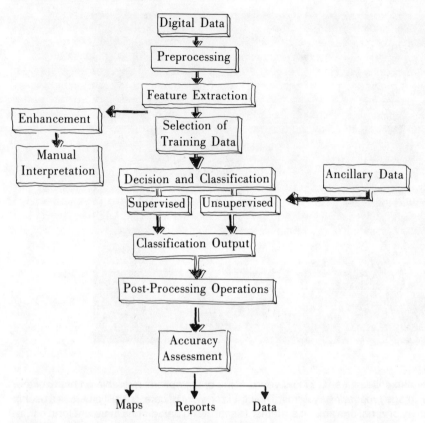

FIGURE 19 SEQUENCE FOR MACHINE INTERPRETATION. This diagram shows an idealized, simplified representation of some of the steps required for digital analysis of remotely sensed data for land use mapping. Actual procedures will vary greatly depending upon specific objectives and circumstances.

based on examination of the covariance structure of several spectral channels, or on measured optical properties of the atmosphere, may be used.

The third group of preprocessing operations deal with *geometric errors,* positional errors in relationships between image representations of ground features and actual geographic relationships between the features as they occur on the earth's surface. Some geometric transformations are not related to errors. For example, *changing of image scale,* and *projection conversion* are operations that simply alter sizes or locational reference systems without consideration of the presence of errors. Many of the usual geometric transformations are necessary because of systematic geometric variations in the movement of the sensor platform or operation of the sensor itself. For example, the simultaneous movement of the Landsat platform and the earth's surface along separate axes cause Landsat data to be skewed relative to correct geographic positions. Geometric preprocessing shifts positions of pixels to approximate true positions on the earth's surface.

Other geometric errors are not directly predictable from a knowledge of sensor operation, but may be caused by interactions between several factors. Often these are not "corrected" (in the usual meaning of the word) but are compensated for by registration of the image to an accurate map, or possibly to another image. There are a variety of methods that have been used to bring one image into registration to another; most require that easily identifiable points common to both images be used to derive linear functions which are then used to derive, from the original image, new data that register to the correct image. The resampled image approximates a hypothetically "correct" image that registers exactly to the true map of the area.

It must be emphasized that all preprocessing procedures alter the original data, perhaps in ways that are not intended or fully understood. Although the objective is, of course, to generate altered data that approximate ideal data, free of error, it seems clear that there are instances in which preprocessing itself introduces undesirable effects, possibly equal in significance to the errors that were removed. There has been very little systematic study of the effects of the usual preprocessing operations upon the performance of the classification algorithms, so users are advised to proceed with caution.

Feature Extraction

Feature extraction reduces the dimensionality of the data. In this context, a 'feature' should be thought of as 'useful information' in the data, rather than a physical feature on the earth's surface. Admittedly, this term is confusing, but its use in this context is well established. Feature extraction serves the practical purpose of reducing the number of variables (in this context, spectral channels) that must be examined, thereby saving time and resources devoted to an analysis. Implicit is the notion that the reduced data set provides information comparable to the complete data set, so that elimination of data results in only minor loss in information. In the ideal, the loss consists mainly of noise present in the raw data. Feature extraction is important in a conceptual sense because it can permit the analyst to focus attention immediately upon the most useful portions of the data by excluding noise and variables containing little additional information.

Subsets of a complete data set form a simple example of feature extraction. The user applies existing information, experience, and the results of previous analyses to select (for example) two of the total of five channels in the data in the belief that the two channels selected provide information comparable in detail to that provided by all five.

Ratios are quotients of values in two channels measured for the same pixel (Tucker 1979). Ratios can reduce the number of channels, and may enhance representations of distributions recorded on the image, if there is an inverse relationship between brightnesses of the same objects in two channels. Feature selection can also be accomplished by other procedures, including principal components analysis, which can reduce several channels of data to a few independent measures that may have great effectiveness in classification.

Image Enhancement

'Image enhancement' has been used with so many different meanings that it is difficult to assign a clear definition. In general, it refers to operations that improve the visual representations of selected distributions on an image for the purpose of sub-sequent manual interpretation. In contrast, feature selection is perhaps most often used when numerical processing is to follow. Image enhancement is often used to prepare images to be used in the search for geologic faults or lineaments not easily recognized in the raw image. Typical enhancement procedures might include ratio-ing, threshold-ing (assigning digital values above or below a certain threshold to a set value), and contrast stretching (altering the range of brightness values to improve the representa-tions of specific features). Many more enhancement operations are available; some are useful in preparing data for quantitative analysis; others may be useful only for visual analysis. Figure 19 shows enhancement as leading to subsequent manual interpreta-tion (Chapter 5), although other options are possible. More than many other proce-dures, enhancement lacks a firm theoretical basis. Many procedures are applied without an understanding of why they are effective, and often their effectiveness varies greatly from one context to another.

Decision and Classification

There are several alternative approaches to the decision and classification step that operates upon the digital data. In this step, the data that have passed through preprocessing and feature extraction or image enhancement are examined and as-signed to informational categories. Spectral values are assigned to the land cover categories of interest to the user.

Supervised classification includes operations that use information derived from a few areas of known identity to classify the remainder of the image. The operator must carefully locate image areas that are known faithfully to represent classes on the final map (Joyce, 1978), based on ground, map, or photographic evidence. These areas must be carefully selected as accurate samples of the larger areas they represent, and their location must be carefully recorded and located on the image. Once these 'training areas,' or 'training fields,' have been accurately located on the image, the classification algorithm uses the spectral means, variances, covariances, and ranges of values within the training fields as a basis for the assignment of pixels of unknown identity to the categories represented by the training data. The extent to which the training areas represent the categories to be mapped determines the accuracy of the final classifica-tion (Hixson 1980; Campbell 1981).

Unsupervised classification, an alternative to supervised classficiation, desig-nates a class of numerical operations that search for 'natural' groupings of the spectral properties of pixels, as examined in multispectral data space. The image analyst then attempts to assign these 'natural' groupings to the informational classes of interest to the map users. The operator who uses supervised classification has a greater degree of interaction with the classification algorithm, and may be able to shape the character and accuracy of the output. Unsupervised classification may reveal the fundamental character of an image, including the numbers of, and interrelationships between, spectral classes. The operator must inspect the output very carefully to determine their correspondence (if any) to informational classes of interest to the user. Although the operator has control over minimum and maximum numbers of categories to be pro-duced by unsupervised classification, he has no control over their identities — a practical disadvantage, especially in situations demanding a specific set of informa-tional classes.

Unsupervised classification requires selection of a few values of initial centers of the spectral clusters that are formed by classification algorithms. Sometimes these starting values are randomly selected to assure a representative sampling of the entire scene. In other instances, starting values are selected in a manner analogous to selection of training data for supervised classification, to represent specific areas known to the operator. Therefore, Figure 19 shows the selection of training data as a common preliminary step for either supervised or unsupervised classification, even though training data serve separate purposes in each of the two approaches.

Although the distinctions between supervised and unsupervised classification may seem quite distinct, there have been advantages in combinations of the two methods. For example, unsupervised classification can be used to define spectrally distinct categories, which can then be used, with other information, as the basis for defining training areas for supervised classification. Alternatively, specific training areas can be selected to form the starting values for forming clusters for unsupervised classification. In general, the most appropriate classification strategy depends upon the basis of the information at hand, experience of the operator, objectives of the study, and characteristics of the region to be mapped (Merchant 1983).

Classification Maps

Classification maps produced by machine are, in respect to logical organization, essentially similar to those produced by manual interpretation, although the two kinds of maps may differ in appearance. The digital format may result in a rather blocky appearance for boundaries on the machine classification (see Wray [1983] for a discussion of a method of presentation that eliminates the blocky appearance). The colors and symbols used to represent categories on the machine classification may differ from those usually used for manually produced maps. These visually differences from traditional maps often are not important, although other differences are significant. If supervised classification has been used, the analyst knows the identities of the colors and symbols of virtue of participation in the process of selecting training data during the classification process. For unsupervised classification, the classification algorithm has control of the identities, and sometimes the numbers, of categories present in the final classification. For example, a single category on the classification map may correspond to two or more spectrally similar land use categories; it may not clearly match any land use category. This general problem of matching spectral categories with informational categories is often a serious issue in machine classification, and is discussed in greater detail below.

Manually produced land use maps usually show relatively large parcels, due in part to the interpreter's generalization of the land use pattern during compilation. Machine classifications, in contrast, may produce patterns characterized by many small parcels scattered in a salt-and-pepper fashion over the mapped area. Such patterns may be caused by the natural landscape pattern, or sometimes by the fact that many classifier algorithms operate upon each pixel in isolation from its neighbors. For the user, the abundance of tiny parcels may constitute a kind of 'noise' that detracts from the usefulness of the map. As a result, the analyst may wish to conduct post-processing operations to produce a map closer in appearance to that produced by manual methods. For example, an algorithm may be designed to search through the map (in its digital form) to identify these small parcels, then merge them with larger surrounding parcels to produce a visually simplified map with larger parcels (Todd *et al.* 1980). The procedure that Thomas (1980) devised considers reclassification based upon spectral properties of pixels, as well as proximity to nearby parcels of contrasting categories. Some post-processing operations may serve essentially cosmetic purposes in that they alter the appearance of the map without necessarily improving accuracy. However, it must be remembered that most manually produced maps include cartograhpic generalizations that can perhaps be considered to have effects analogous to post-processing operations.

Machine classifications of land use and land cover patterns date from the early applications of digital classification of remotely sensed data, so there is a large body of knowledge and experience that can be applied to problems in land use mapping. Furthermore, some of the most interesting research in digital classification has been conducted by those who have attempted to solve many of the problems encountered in the mapping of land cover. The following sections outline some topics of special interest for those working with automated classification of land use.

Assigning Spectral Signatures to Information Categories

Those who use the final products of a land use inventory have an interest in informational categories, such as 'urban,' 'agricultural,' or 'forested' land. These categories are so named because they convey knowledge of the subject at hand, rather than the brightness and spectral information conveyed by the raw data from the sensor. Informational categories can be derived from remotely sensed images to the extent that they can be consistently and accurately associated with spectral and brightness information on the image, using the characteristic colors, brightnesses, and textures that may characterize specific categories. The term 'spectral signature' refers to spectral responses of specific informational categories, although the term implies a consistency and precision that is seldom present.

Thus, a fundamental problem in the practice of remote sensing is to match brightness and spectral categories on the image to the informational categories of interest. Analysts accomplish this matching using a complex association of the elements of image interpretation, experience, and contextual information. During automated interpretation, the matching process must be conducted in a much more abstract manner, using rather limited information relative to that available to a human interpreter.

Informational categories seldom correspond on a one-to-one basis to specific spectral categories. For example, a given informational category, such as forested land, will typically manifest itself as several spectral categories due to natural variations in density, maturity, and species composition and due also to variations in illumination, slope, aspect, and other factors. As a result, "forest" can be mapped only by finding and identifying a variety of diverse spectral patterns related to these variations in composition and illumination. Figure 20 illustrates multispectral data for a natural scene in which informational categories appear as several distinct spectral categories. One of the basic tasks of image analysts is to guide the classification algorithm through the process of finding, then identifying, the diverse spectral patterns on the image.

Here we can provide little specific advice on this subject except to emphasize that the task requires an intimate knowledge of the subject, the geographic region, the

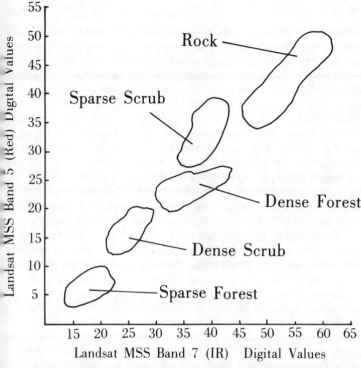

FIGURE 20   SPECTRAL CATEGORIES. This diagram represents remotely sensed data from two separate spectral channels plotted on a single diagram. Landsat multispectral scanner band 5 is represented by the vertical axis; band 7 is shown along the horizontal axis. Brightness increases from bottom to top and from left to right. Because brightness of objects tend to increase simultaneously in both bands, data for the scene tend to group together along a line oriented at about 45 degrees from the origin. The areas delineated here represent relatively homogeneous spectral regions as defined by unsupervised classification. Unsupervised classification procedures cannot, however, assign these spectral regions to the informational categories of interest to the map user. The analyst must accurately label each spectral category. Note that a single informational class ("forest") may correspond to several spectral regions, and that not all spectral categories neatly match the usual land use categories — the problem of matching spectral classes to informational classes. Because spectral properties change with time, the identifications for this scene will not necessarily apply to others.

image, and the operation of the classification algorithm. In this context, it should be clear that use of an automated classification procedure cannot replace the knowledge and skill required of the image interpreter who uses manual techniques.

Textural Classifiers

Land use classification often presents special problems for digital analysis because of the diverse spectral characteristics of many land use categories. Idealized land use regions consist of spectrally homogeneous patches on the earth's surface. Accurate mapping of such regions could be accomplished by the relatively straightforward process of matching spectral categories to the spectral 'signatures' of informational categories.

Actual land use regions are, of course, formed often by assemblages of spectrally diverse features. For example, low density residential land, as viewed from above at relatively fine spatial resolution, is composed largely of tree crowns, rooftops, lawns, and paved street, driveways, and parking lots. Although human interpreters can readily recognize such complex patterns, many digital classification algorithms encounter serious problems in accurate classification of such scenes because they are designed to classify each separate spectral region as a separate informational category. In reality such categories are defined by the entire collection of diverse spectral responses.

Textural classifiers attempt to measure 'image texture,' distinctive spatial and spectral relationships between neighboring pixels. For example, the standard deviation of brightness values within a neighborhood of specified size, systematically positioned over the entire image, may provide a rough measure of the spectral variability over short distances as a measure of image texture. Such a measure may permit the analyst to classify composite categories such as the one mentioned above (Jensen and Toll 1982). Usually, more sophisticated measures of texture are required to produce satisfactory results. For example, other textural measures examine relationships between brightness values at varying distances and directions from a central pixel, which is systematically moved over the image (Maurer 1974; Haralick et al. 1973; Haralick 1979).

Jensen (1979) found the use of textural measures improved his classification of Level II and III suburban and transitional land using band 5 of the Landsat multispectral scanner. Improvements were confined to certain categories and were, perhaps, of marginal significance when considered in the context of increased cost.

Textural measures seem to work best when relatively large neighborhoods are defined (perhaps as large as 64 by 64 pixels). Such large neighborhoods may cause problems when they straddle boundaries between categories. In addition, such large neighborhoods may decrease the effective spatial resolution of the final map, as they must form the smallest spatial elements on the map. For these reasons large neighborhoods can be considered effective only when very large areas are examined.

Ancillary Data

'Ancillary data' refers to the use of non-image information as an aid for classification of spectral data (Tom et al. 1978). In the process of manual interpretation of images, there is a long tradition of both implicit and explicit use of such information, including data from maps, photographs, field observations, reports, and personal experience. For digital analysis, ancillary data often consist of data available in formats consistent with the digital spectral data, or in forms that can be conveniently transformed into usable formats. Examples include digital elevation data, or digitized soil maps (Anuta 1976).

Ancillary data can be used in either of two ways. They can be 'added' to, or 'superimposed' over, the spectral data to form a single multiband image; the ancillary data are treated simply as additional channels of information. Alternatively, the analysis can proceed in two steps using a layered classification strategy (as described below); the spectral data are classified in the first step, then the ancillary data form the basis for reclassification and refinement of the initial results.

A serious limitation to practical application of digital ancillary data is the problem of incompatibility with the remotely sensed data. Physical compatibility (such as matching digital formats) has, of course, obvious practical significance. Logical compatibility may form a more subtle, but equally important problem. Seldom are ancillary data collected specifically for use with a specific remote sensing problem; usually as a means of reducing costs in time and money, they are derived from archival data collected for another purpose. For example, digital terrain data gathered by the U.S. Geological Survey and by the Defense Mapping Agency for topographic quadrangles are frequently used as ancillary data for remote sensing studies. Seldom, if ever, could remote sensing projects absorb the costs of digitizing, editing, and correcting these data for specific use in a single study. One consequence of this practice is that the ancillary data are seldom compatible with the remotely sensed data in scale, resolution, date, and accuracy. Some differences can be minimized by preprocessing of the ancillary data to reduce effects of different measurement scales, resolutions, and the like. In other situations, unresolved incompatibilities, possibly quite subtle, may detract from the potential effectiveness of the ancillary data.

Choice of ancillary variables may be critical. In the mountainous regions of the western U.S., elevation data have been very effective as ancillary data for mapping vegetation patterns with digital Landsat data, due in part to the large ranges in local elevation, and the close associations of vegetation distributions with elevation, slope, and aspect. In other settings where elevation differences may have more subtle influences upon vegetation distributions, such data may not form effective ancillary variables. Although some scientists advocate use of all available ancillary data, in the hope of deriving whatever advantage might be possible, common sense would seem to favor careful selection of those variables with conceptual and practical significance to the mapped distributions.

Layered Classifications

'Layered classification' refers to use of a hierarchial process in which two or more steps form the basis for classification. The usual single-step classification algorithm uses all available information in a single process to produce the entire menu of informational categories. Layered classification uses subsets of data in a series of separate steps, presumably by applying each form of information in its most effective context. The hierarchical structure permits the most difficult classification decisions to be made in a context that isolates the problem categories from others and that focuses the most effective variables upon that classification decision.

For example, Jensen (1978) devised a layered classification strategy that separated vegetated and non-vegetated regions early in the classification by using vegetation indices (analogous to those discussed by Tucker 1979) as a measure of extent of living biomass within each pixel. This basic distinction separated built-up and heavily vegetated regions at an early point in the classification, preventing confusion between categories later in the classification process.

Layered classification can be useful only if the classification logic can be structured in a way that minimizes errors at the upper levels of the decision tree. If errors are made at the first stage, they are carried to lower levels, and will appear in the final product regardless of the soundness of subsequent decisions.

Contextual Classifiers

Contextual information is that information derived from spatial relationships among pixels within a given image. Whereas textural usually refers to spatial interrelationships among unclassified pixels within a window of specified size, context is determined by positional relationships between pixels, either classified or unclassified, anywhere within the scene (Swain *et al.* 1981; Gurney and Townshend 1983).

Contextual classifiers are efforts to simulate some of the higher-order interpretation processes used by human interpreters, in which the identity of an image region is derived, in part, from its location in relation to other regions of specified identity. For example, a human interpreter considers sizes and shapes of parcels in identifying land use, as well as the identities of neighboring parcels. The characteristic spatial arrangement of commercial, industrial, residential, and agricultural land in an urban region permits the interpeter to identify parcels that might be indistinct if considered in respect to spectral properties of individual pixels.

Contextual classifiers can also operate upon classified data to reclassify erroneously classified pixels, or to reclassify isolated pixels (perhaps correctly classified) that form regions so small and so isolated that they are of little interest to the user. Such uses may be considered to be essentially cosmetic operations, but they are useful in editing the results for final presentation.

7

Accuracy Assessment

Study of the accuracy of maps and reports based upon analysis of remotely sensed data has many implications, including their legal standing, usefulness for management of public and private lands, and validity for scientific research. Sound administrative and management decisions rest, of course, on the accuracy of available information. Administrators must know the accuracy of data at hand and the relative value of data from different sources. If their decisions are challenged in court, data validity becomes a key issue, especially when information is derived from satellite imagery, or from abstract analyses not immediately obvious to the layman. Thus, when data sources and analytical methods are unfamiliar to the public, concrete evidence of accuracy is required to establish credibility.

From a scientific perspective, analysis of the comparative accuracies of alternative approaches to image interpretation have great significance for the use of remotely sensed data in many fields. For example, there have been few systematic investigations of the relative accuracies of manual and machine interpretations, accuracies of different individuals, accuracies of the same interpreter at different times, accuracies associated with separate preprocessing and classification algorithms, or accuracies associated with different images of the same area.

The accuracy of information generated from remotely sensed data has been of interest for many years, but recent widespread evaluation of the results of digital image classification has probably been responsible for the major growth in accuracy assessment. In a way, the focus upon accuracy of digital classifications is unfair, because traditionally the usefulness of manual interpretations has often been accepted in the absence of substantive evidence of accuracy. As a result, it is easy to conclude that recent interest in accuracy assessment has its source in a suspicion of the quality of many digital classifications, rather than in a pure interest in accuracy for its own sake. In any event, we should not feel obligated to accept at face value the quality of any map, regardless of its origin or appearance, without supporting evidence. Generating the kind of data necessary to support credible statements of map accuracy is a very complex subject that has generated a rather sophisticated debate concerning the most appropriate methods to be used. We can only introduce some of the basic issues; the reader should refer to the references cited for more complete discussions.

Definition and Measurement of Map Accuracy

In most instances the accuracy assessment problem is essentially one of comparing two maps, one based upon analysis of remotely sensed data (usually the map to be tested, or evaluated) and another based upon a different source of information (usually considered to be the standard for the comparison). Both maps are assumed to be in the traditional form of most land use maps; that is, both are formed by subdividing a geographic region into a mosaic of discrete, labeled parcels. To assess the accuracy of the first map, it is necessary that the two maps register to one another, and that the two maps use classification systems comparable in number of categories and mapping detail.

The least complicated, oldest, and, possibly, most frequently used method of evaluation is simply to compare the two maps in respect to the areas that match when the two maps are superimposed. The result of such a comparison is to report the areal proportions of the two patterns that match ("50 percent," "97 percent"). These values report the extent of the agreement between the two maps in total area in each category, but do not take into account compensating errors in misclassification that can cause this kind of accuracy measure to itself be inaccurate. This form of error assessment is sometimes refered to as 'non-site-specific accuracy' because it does not consider agreement between the two maps at specific locations, but only overall agreement.

Non-site-specific accuracy has been a widely used method of reporting accuracies of land use maps. Anderson and colleagues (1976) stated that accuracies of 85 percent are required for satisfactory use of land use data for resource management. Fitzpatrick-Lins (1978) reported that accuracies of USGS land cover maps of the central Atlantic coastal region of the U.S. are about 85 percent (1:24,000), 77 percent (1:100,000), and 73 percent (1:250,000) accurate. For automated interpretations of land use in the Denver metropolitan area (six level I categories), Tom and collaborators (1978) reported accuracies of about 38 percent using only Landsat data (including band ratios), and about 78 percent using ancillary data.

The second form of accuracy measure, 'site-specific accuracy,' is based upon detailed assessment of agreement between the two maps at specific locations on the maps. In most analyses, the units of comparisons are defined as pixel-sized units from the map derived from remotely sensed data, although, if necessary, any pair of matching maps could be compared by using any network of uniform cells to define the units of comparison. Site-specific accuracy has been measured using several alternative strategies. For supervised classification, the simplest strategy is to compare the classified data with the training samples used to generate the classification. If training samples have been positioned at random throughout the mapped area, they can perhaps be considered to form representative sample of the scene as a whole. It is, however, difficult to have confidence in such a procedure because the number of training samples is often small in relation to the numbers of pixels to be classified. Furthermore, it seems inevitable that accuracies for training data are likely to be higher than for the map as a whole. In fact, for many classification algorithms, one would expect correct classification of all training data if the classification has been conducted in accord with its statistical foundations. Strictly speaking, then, any result short of 100 percent accuracy detracts from the credibility of the entire map. It would seem that this procedure is born of expediency and, as a result, can have little use in any serious attempt at accuracy assessment.

At a minimum, accuracy evaluation should be based upon sites of known identity *not* used in the training data. This procedure yields a much more credible evaluation, but presents the practical problem of devoting time and resources to collection of the required data (data that will not themselves improve accuracy) as well as scientific problems of deciding how many such sites are required and where they should be located. A further alternative strategy is based on sampling the entire scene (perhaps 2 to 5 percent of pixels in a complete Landsat scene, but probably more for smaller regions). Aside from the practical problems of acquiring data in remote areas, important research questions concern strategies for selecting and classifying samples, continuation design of appropriate non-parametric statistical tests, and the methods of best reporting results to permit comparison with other studies. For example, should samples be random, clustered, or stratified?

The standard form for reporting site-specific error is the error matrix, sometimes referred to as a 'confusion matrix' because it identifies not only overall errors for each category, but also misclassifications, due to confusion between categories. The error matrix is essential for any serious study of accuracy. The error matrix consists of an **n** by **n** array, where **n** is equal to the number of categories on the map (Table 11). The upper edge is labeled with the categories on the reference ("correct") classification. The left hand side is labeled with the same **n** categories; these refer to locations on the map being evaluated (The meanings of the two axes are reversed in some applications, but the interpretation of the matrix is not altered). Diagonal entries in the matrix reveal the number of cells or pixels (sometimes the percentages) in a given category that were *correctly* classified on the map to be evaluated.

Examination of the error matrix reveals errors of omission and errors of commission for each category. Errors of omission are, for example, the assignment of areas of forest on the ground to the agricultural category on the map (an area of 'real' forest on the ground has been omitted from the map). Using the same example, an error of commission would be to assign an area of agriculture on the ground to the forest category on the map. This distinction is essential because an interpretation could achieve 100 percent accuracy in respect to forest by assigning *all* pixels to forest. The

TABLE 11   EXAMPLE OF AN ERROR MATRIX[a]

Actual Land Use	Interpreted Land Use (Number of Pixels)						
	Urban	Agriculture	Range	Forest	Water	Barren	Total
Urban	180,781	13,563	6,287	175	1,292	465	202,563
Agriculture	29,299	46,619	5,296	0	826	157	82,197
Range	4,005	5,568	15,788	307	53	1,923	28,044
Forest	246	462	780	3,328	2	204	5,022
Water	1,830	667	72	0	4,190	0	6,759
Barren	3	3	' 4,248	424	0	2,513	7,191
Total	216,164	67,282	32,471	4,234	6,363	5,262	216,164

[a]Columns list classification results; rows show correct ground data.
Source: Compiled from data presented by Tom *et al.* 1978.

tabulations of errors of commission reveal such achievements to be meaningless, because there will be high errors of commission for forest, and high errors of omission for other categories.

Interpreting The Error Matrix

Table 11 shows an error matrix generated by a machine classification of a 576 by 576 segment of a Landsat scene (Tom *et al.* 1978). Each of the 331,776 pixels in this segment was assigned to one of six land cover categories. The resulting classification was then compared, pixel by pixel, to a previously existing land use map of the same area, and the differences were tabulated, category by category, to form the data for Table 11.

The column on the right edge of the matrix gives total pixels in each category on the original land use map; the row at the bottom shows total pixels in each category in the classified scene. The diagonal entries show number of correctly classified pixels — range land classified as range, urban land classified as urban. The diagonal sequence of entries from upper left to lower right splits the matrix into two halves. The upper left half is formed by values derived from the classification map, tabulated by their correct identities from the labeled use map. The lower left half is formed by values derived from the land use map, tabulated by their assignment on the classification map.

Thus, the off-diagonal entries show errors of omission and commission. The upper right half of the matrix tabulates errors of omission, such as urban land on the ground classified as range, or agricultural land classified as urban. Entries that form the lower left half of the matrix reveal errors of commission. For example, land other than urban has been classified as urban, especially agricultural land (29,299) and range (4,005). Non-diagonal row entries sum to give total errors of omission; non-diagonal column entries sum to give total errors of commission.

Table 12 summarizes errors of omission and commission by land use category. For example, there were 216,164 pixels of urban land in the land use map; of these, 180,781 were correctly classified as urban land, about 84%. The remaining 35,383 pixels (the sum of the off-diagonal entries from column 1, Table 11), forming about 16 percent of the total urban land, were incorrectly classified, mainly as agricultural land, but also as other categories. These form the errors of commission for urban land.

Errors of omission (equal to about 11 percent of the total of urban land in the area) consist mainly of urban land classified as agricultural land and rangeland, as is shown in Table 11. For this classification of this particular scene, it is evident that confusion of urban and agricultural land is a major source of error in the classification of urban land use.

For a contrasting relationship between errors of omission and commission, note that urban land tends to be classified as barren land, whereas very seldom is barren land classified as urban. Inspection of the error matrix reveals the kinds of errors generated by the classification process, which may in turn permit improved interpretation of the map and improve accuracy in future classifications.

There is considerable interest in comparing error matrices generated by different classification processes, different interpreters, and different forms of imagery (to name only a few of the many forms of comparison that could be examined). This form of analysis depends upon standardization of error matrices to bring totals for all matrices to the same value. This step is required so that the analysis can examine the pattern of errors within each matrix without regard to differences caused by the total number of

TABLE 12   SUMMARY OF COMMISSION/OMISSION ERRORS REPORTED IN TABLE 11.

Category	Omission Errors		Commission Errors		Correct	
	Total[a]	Percent	Total	Percent	Total	Percent
Urban	21,782	11	35,383	16	180,781	84
Agriculture	35,578	43	20,663	31	46,619	69
Range	12,156	44	16,683	51	15,788	49
Forest	1,694	34	906	21	3,328	79
Water	2,569	38	2,173	34	4,190	66
Barren	4,678	65	2,749	52	2,513	48

[a]Totals = number of pixels.
Source: Compiled from data presented by Tom *et al.* 1978.

pixels in the various scenes. Standardization is usually accomplished by a process that brings all row and column totals to a common value, usually 1. Comparisons of standardized matrices can then be conducted using one of several alternative statistical tests selected specifically for their appropriateness in this context. Details can be found in Congalton and Mead (1983).

Sources of Classification Error

Classification error in machine analysis of remotely sensed data results from complex interactions between the spatial structure of the landscape, sensor resolution, preprocessing algorithms, and classification procedures. Perhaps the simplest causes of error are related to the misassignment of informational categories to spectral categories. The bare granite of mountainous areas, for example, can be easily confused with the spectral response of concrete in urban areas. However, the sources of most errors in machine classifications are probably more complex. Mixed pixels occur as resolution elements of a remote sensing system fall on the boundaries between separate land use parcels. These pixels may well have digital values unlike either of the two categories, and may easily be misclassified even by the most accurate and robust classification procedures. Such errors are often visible in digital classification products as chains of misclassified pixels that parallel the borders of rather large, homogeneous, parcels.

It is in this manner that the character of the landscape contributes to the potential for error through the complex patterns of parcels that form the scene that is imaged. A very simple landscape composed of large, uniform, distinct categories is likely to be easier to classify accurately than one with small, heterogeneous, indistinct parcels arranged in a complex pattern. Key landscape variables are likely to include:

(1) parcel size;
(2) variation in parcel size;
(3) parcel identities;
(4) numbers of categories present;
(5) arrangement of parcels;
(6) number of parcels per category;
(7) shapes of parcels; and
(8) radiometric and spectral contrast with surrounding parcels

These variables change from one region to another (Simonett and Coiner 1971 Podwysocki 1976) and, within a given region, from season to season. As a result, the sources of error in a given image are not necessarily predictable from previous experience in other regions or at other dates.

The sensor itself may introduce errors through its interaction, at given levels of spatial, radiometric, and spectral resolution, with the landscape variables mentioned above. In addition, noise and spatial dependencies may be introduced by sensor design and operation. Preprocessing operations designed to correct radiometric and geometric errors may themselves introduce characteristics that may lead to subsequent errors in the classification process. For example, resampling for geometric corrections may introduce spatial dependencies not originally present in the raw data. And the classification process itself may interact with original or introduced characteristics of the data to yield subtle classification errors in the final product. In brief, classification errors should not be considered to be the inevitable result of undefinable causes. They are, in part, the result of properties of the scene that has been imaged, the design and performance of the sensor, and of preprocessing procedures.

Error Characteristics

Classification errors are assignments of pixels belonging to one category (as determined by ground observation) to another category based upon remotely sensed data. There are few systematic studies of geographic characteristics of these errors, but experience and logic suggest that errors are likely to possess at least some of the characteristics listed:

(1) Errors are not distributed over the image at random, but, to a degree, display systematic, ordered occurrence in space;

(2) Errors are not distributed at random among the various categories on the image, but may be preferentially associated with certain categories;

(3) Erroneously assigned pixels are often not spatially isolated, but occur grouped in areas of varied size and shape (Campbell 1981);

(4) Errors are not distributed randomly among the various parcels on the image, but may occur in parcels with certain sizes, shapes, locations, and arrangement in respect to other parcels; and

(5) Errors may have specific spatial relationships within the parcels to which they pertain; for example, they may tend to occur at the edges or in interiors of parcels.

Accuracy Assessment Research

The validity of accuracy estimates based on sampling depends upon: (a) the number of samples, (b) the arrangement (pattern) of samples over the image, (c) the spacing between samples in a given sampling pattern, and (d) the allocation of samples of the several categories on the map. In general, accuracy studies have focused primarily on these issues, and have devoted only passing attention to placement of samples on the image. These neglected considerations are of great practical and theoretical significance, as even cursory examination of error patterns suggest that many errors are not distributed at random and may escape detection by the usual sampling strategies.

Although the essential elements of the accuracy assessment problem had been outlined earlier, the work by Hord and Brooner (1976) forms a starting point for much of the current research in accuracy assessment. They recognized that observed overall accuracy is the result of some combination of misclassified polygons, errors in boundary placement, and control point placement error. They applied sampling patterns to examine polygon classification errors. Their work explicitly recognized the probabilistic character of the problem, and recommended that accuracy be reported with confidence limits. Van Gendren and Lock (1978) studied land cover map accuracy using a stratified random sample; they recommended that scientists choose the sample size required to justify a desired level of confidence. Ginevan (1979) reviewed earlier research, concluding that accuracy studies should be designed simultaneously to minimize sample size and minimize the probability of accepting an erroneous classification or rejecting a correct classification. Chrisman (1980) rejected the use of correct percentage as an indication of map accuracy, because under certain circumstances even a random classifier can produce ostensibly reasonable values. He suggested the use of an index that reports the agreement between two maps in relation to results expected from a random assignment of pixels to categories. For example, he recommended *Cohen's Kappa,* a statistic that varies from +1 (perfect agreement) to -1 (complete disagreement). A value of zero indicates that the results are not distinguishable from those obtained by a random classification. Current research in this field seems to be based upon measures of this kind.

Key issues that should be addressed in future accuracy research include definition of strategies for application of non-parametric statistics to accuracy assessment, design of optimum sampling strategies, and the identification of statistical and geographic properties of classification errors. Of particular interest are topics that have been peripheral to the research mentioned above, especially those requiring a knowledge of the spatial properties of errors, and their influence upon the accuracy measures now in use.

Bibliography

Abel, E. 1966. *The Missile Crisis*. New York: J.P. Lippincott.

Anderson, J.R. 1961. "Toward More Effective Methods of Obtaining Land Use Data in Geographic Research," *The Professional Geographer* 13:15-18.

Anderson, J.R. 1971. "Land-Use Classification Schemes," *Photogrammetric Engineering* 37:379-387.

Anderson, J.R. et al. 1976. *A Land Use and Land Cover Classification System for Use with Remote Sensor Data*. U.S. Geological Survey Professional Paper 964. Washington, DC: USGPO.

Anderson, J.R. 1977. "Land Use and Land Cover Changes — A Framework for Monitoring," *Journal of Research, U.S. Geological Survey* 5, 3:143-153.

Anderson, J.R. (editor). 1977. "Land Use and Land Cover Maps and Statistics from Remotely Sensed Data," *Remote Sensing of the Electromagnetic Spectrum* 4, 4.

Anuta, P.E. 1976. "Digital Registration of Topographic Data and Satellite MSS Data for Augmented Spectral Analysis," pp. 180-187 in *Proceedings, 42nd Annual Meeting, American Society of Photogrammetry*. Falls Church, VA: American Society of Photogrammetry.

Avery, T.E. 1977. *Interpretation of Aerial Photographs*. Minneapolis, MN: Burgess.

Babington Smith, C. 1974. *Evidence in Camera: The Story of Photographic Intelligence in World War II*. London: David and Charles (originally published as *Air Spy*, 1957).

Baker, R.D. et al. 1979. "Land Use/Land-Cover Mapping From Aerial Photographs," *Photogrammetric Engineering and Remote Sensing* 45:661-668.

Baker, O.E. 1926. "Agricultural Regions of North America," *Economic Geography* 2:459-489 (one of several in a series of articles).

Borchert, J.R. et al. 1974. *Perspective on Minnesota Land Use: 1974*. Minneapolis, MN: State Planning Agency.

Boulaine, J. 1980. *Pédologie Appliquée*. Paris: Masson.

Brown, R.A. 1968. *The Normans and the Norman Conquest*. New York: Thomas Crowell.

Bryan, M.L. 1982. "Urban Land Use Classification Using Synthetic Aperture Radar," pp. 124-134 in Richason 1982.

Burley, T.M. 1961. "Land Use or Lan Utilization?" *The Professional Geographer* 13, 6:18-20.

Campbell, J.B. 1978. "A Geographical Analysis of Image Interpretation Methods," *The Professional Geographer* 30:264-269.

Campbell, J.B. 1981. "Spatial Correlation Effects Upon Accuracy of Supervised Classification of Land Cover," *Photogrammetric Engineering and Remote Sensing* 47:355-363.

Carter, V. et al. 1977. *Summary Tables for Selected Digital Image Processing Systems*. U.S. Geological Survey Open File Report 77-414. Reston, VA: USGS

Chrisman, N.R. 1980. "Assessing Landsat Accuracy: A Geographic Application of Misclassification Analysis," Second Colloquium on Quantitative and Theoretical Geography. Cambridge, UK.

Christian, C.S. 1959. "The Eco-Complex and Its Importance for Agricultural Assessment," Chapter 36, Biogeography and Ecology in Australia, in *Monographiae Biologicae* 8:587-605.

Christodoulou, D. 1959. *The Evolution of the Rural Land Use Pattern in Cyprus*. Bude, UK: Geographical Publications Ltd. (The World Land Use Survey, Monograph 2).

Clawson, M. et al. 1960. *Land For the Future*. Baltimore, MD: Johns Hopkins Press.

Coiner, J.C. and S.A. Morain. 1971. "Image Interpretation Keys to Support Analysis of SLAR Imagery," pp. 393-412 in *Proceedings, Fall Technical Meeting of the American Society of Photogrammetry.* Falls Church, VA: American Society of Photogrammetry.

Coleman, A. 1961. "The Second Land Use Survey: Progress and Prospect," *Geographical Journal* 127:168-186.

Coleman, A. 1964. "Some Cartographic Aspects of the Second Series Land Use Maps," *Geographical Journal* 130:167-170.

Coleman, A. 1980. "Land Use Survey Today and Tomorrow," pp. 216-228 in E. H. Brown (editor), *Geography: Yesterday and Tomorrow.* New York: Oxford University Press.

Colwell, R.N. (editor). 1960. *Manual of Photographic Interpretation.* Falls Church, VA: American Society of Photogrammetry.

Congalton, R.G. and R.A. Mead. 1983. "A Quantitative Method to Test for Consistency and Correctness in Photointerpretation," *Photogrammetric Engineering and Remote Sensing* 49:69-74.

Darby, H.C. 1952. *The Doomsday Geography of Eastern England.* New York: Cambridge University Press (the first of a series of volumes).

Darby, H.C. 1977. *Domesday England.* New York: Cambridge University Press.

de Neufville, J.I. (editor). 1981. *The Land Use Policy Debate in the United States.* New York: Plenum.

Dickinson, G.C., and M.G. Shaw. 1977. "What is 'Land Use'?" *Area* 9:38-42.

Drake, B. 1977. "Necessity to Adapt Land Use and Land Cover Classification Systems to Readily Accept Radar Data," pp. 993-1000 in *Proceedings, 11th International Symposium on Remote Sensing of Environment.* Ann Arbor, MI: ERIM.

Espenshade, E.B. (editor) 1960. *Goode's World Atlas.* Chicago: Rand McNally.

Estes, J.E. and L. Senger. 1972. "Remote Sensing in the Detection of Regional Change," pp. 317-324 in *Proceedings, 8th International Symposium on Remote Sensing of Environment.* Ann Arbor, MI: ERIM.

Estes, J.E. 1982. "Remote Sensing and Geographic Information Systems Coming of Age in the Eighties," pp. 23-40 in Richason 1982.

Estes, J.E. *et al.* 1982. "Monitoring Land Use and Land Cover Changes," pp. 100-110 in Johannsen and Sanders 1982.

Estes, J.E. and D.S. Simonett. 1975. "Fundamentals of Image Interpretation," pp. 869-1076 in R.G. Reeves (editor), *Manual of Remote Sensing.* Falls Church, VA: American Society of Photogrammetry.

Finn, R.W. 1961. *The Domesday Inquest and the Making of Domesday Book.* Westport, CT: Greenwood Press.

Fitzpatrick-Lins, K. 1978. "An Evaluation of Errors in Mapping Land Use Changes for the Central Atlantic Regional Ecological Test Site," *Journal of Research, U.S. Geological Survey* 6:339-346.

Fridland, V.M. 1972. *Pattern of the Soil Cover.* Moscow (Israel Program for Scientific Translation, 1976).

Galbraith, V.H. 1961. *The Making of Domesday Book.* Oxford, UK: Clarendon Press.

Ginevan, M.E. 1979. "Testing Land-Use Map Accuracy: Another Look," *Photogrammetric Engineering and Remote Sensing* 45:1371-1377.

Goddard, G.W. and D.S. Copp. 1969. *Overview: A Lifelong Adventure in Aerial Photography.* Garden City, NY: Doubleday.

Gurney, C.M. and J.R. Townshend. 1983. "The Use of Contextual Information in the Classification of Remotely Sensed Data," *Photogrammetric Engineering and Remote Sensing* 49:55-64.

Gutkind, A.E. 1956. "Our World From the Air: Conflict and Adaptation," pp. 1-33 in W.L. Thomas (editor), *Man's Role in Changing the Face of the Earth.* Chicago: University of Chicago Press.

Haralick, R.M. *et al.* 1973. "Textural Features for Image Classification," *IEEE Transactions on Systems, Man, and Cybernetics* SMC-3:610-622.

Haralick, R.M. 1979. "Statistical and Structural Approaches to Texture." *Proceedings of the IEEE* 67:786-804.

Hardy, E.E. 1970. "Inventorying New York's Land Use and Natural Resources," *New York's Food and Life Sciences Quarterly* 3, 4:4-7.

Heath, G.R. 1956. "A Comparison of Two Basic Theories of Land Classification and Their Adaptability to Regional Photo Interpretation Key Techniques," *Photogrammetric Engineering* 22:144-168.

Hilsman, R. 1967. *To Move a Nation: The Politics of Foreign Policy in the Administration of John F. Kennedy.* Garden City, NY: Doubleday.

Hixson, M. *et al.* 1980. "Evaluation of Several Schemes for Classification of Remotely Sensed Data," *Photogrammetric Engineering and Remote Sensing* 46:1547-1553.

Hord, R.M., and W. Brooner. 1976. "Land Use Map Accuracy Criteria," *Photogrammetric Engineering* 42:671-677.

Hoskins, W.G. 1955. *The Making of the English Landscape.* Baltimore: Penguin.

Hsu, M.L. *et al.* 1975. "Computer Applications in Land-Use Mapping and the Minnesota Land Management Information System," pp. 298-310 in J.C. Davis and M.J. McCullagh (editors), *Display and Analysis of Spatial Data.* New York: Wiley.

Hudson, G.D. 1936. "The Unit Area Method of Land Classification," *Annals,* Association of American Geographers 26:99-112.

International Geographical Union. 1952. *Report of the Commission on World Land Survey for the Period 1949-1952.* Worcester, MA.

Jackson, R.H. 1981. *Land Use in America.* New York: Wiley.

James, P.E. and C.F. Jones (editors). 1965. *American Geography: Inventory and Prospect.* Syracuse: Syracuse University Press and Association of American Geographers.

Jensen, J.R. 1978. "Digital and Land Cover Mapping Using Layered Classification Logic and Physical Composition Attributes," *The American Cartographer* 5:121-132.

Jensen, J.R. 1979. "Spectral and Textural Features to Classify Elusive Land Cover at the Urban Fringe," *The Professional Geographer* 31:400-409.

Jensen, J.R. and D.L. Toll. 1982. "Detecting Residential Land Use Development at the Urban Fringe," *Photogrammetric Engineering and Remote Sensing* 48:629-643.

Johannsen, C.J. and J.L. Sanders (editors). 1982. *Remote Sensing for Resource Management.* Ankeny, IA: Soil Conservation Society of America.

Joyce, A.T. 1978. *Procedures for Gathering Ground Truth Information for a Supervised Approach to a Computer Implemented Land Cover Classification of Landsat-Acquired Multispectral Scanner Data.* NASA Reference Publication 1015. Houston: National Aeronautics and Space Administration.

Kellogg, C.E. and A.C. Orvedal. 1969. "Potentially Arable Soils of the World and Critical Measures for Their Use," *Advances in Agronomy* 21:109-170.

Landis, G. 1955. "Concept and Validity of Association Photographic Interpretation Keys in Regional Analysis," *Photogrammetric Engineering* 21:705-706.

Lebon, J.H.G. 1965. *Land Use in Sudan.* Bude, UK: Geographical Publications, Ltd. (The World Land Use Survey, Monograph 4).

Lee, W.T. 1922. *The Face of the Earth as Seen From the Air: A Study in the Application of Airplane Photography to Geography.* New York: American Geographical Society, Special Publication 4.

Lillesand, T.M. and R.W. Kiefer. 1979. *Remote Sensing and Image Interpretation.* New York: Wiley

Loelkes, G.L. 1977. *Specifications for Land Cover and Associated Maps.* U.S. Geological Survey Open File Report 77-555. Reston, VA: USGS.

Lounsbury, J.F. and F.T. Aldrich. 1979. *Introduction to Geographic Field Methods.* Columbus: Charles Merrill.

Lounsbury, J.F. et al. 1981. *Land Use: A Spatial Approach.* Dubuque, IA: Kendall/ Hunt.

Marschner, F.J. 1959. *Land Use and Its Patterns in the United States.* Washington, DC: U.S. Department of Agriculture, Agriculture Handbook No. 153 (with map at 1:5,000,000).

Maurer, H. 1974. "Quantification of Textures — Textural Parameters and Their Significance for Classification of Crop Types From Colour Aerial Photographs," *Photogrammetria* 30:21-40.

Merchant, J.W. 1982. "Employing Landsat MSS Data in Land Use Mapping: Observations and Considerations," pp. 71-91 in Richason 1982.

Mitchell, W.B. et al. 1977. *GIRAS: A Geographic Information Retrieval and Analysis System for Handling Land Use and Land Cover Data.* U.S. Geological Survey Professional Paper 1059. Washington, DC: USGPO.

Moik, J.G. 1980. *Digital Processing of Remotely Sensed Images.* NASA SP-431. Washington, DC: USGPO.

Monkhouse, F.J. and Wilkinson, H.R. 1971. *Maps and Diagrams.* London: Methuen.

National Academy of Sciences. 1977. *Resource Sensing From Space: Prospects For Developing Countries.* Washington, DC.

Niddrie, D.L. 1961. *Land Use and Population in Tobago: An Environmental Study.* Bude, UK: Geographical Publications, Ltd. (The World Land Use Survey, Monograph 3).

Nunnally, N.R. and R.E. Witmer. 1970. "Remote Sensing for Land-Use Studies," *Photogrammetric Engineering* 36:449-453.

Peplies, R.W. and H.F. Keuper. 1975. "Regional Analysis," pp. 1947-1998 in R.G. Reeves (editor), *Manual of Remte Sensing.* Falls Church, VA: American Society of Photogrammetry.

Podwysocki, M.H. 1976. *An Estimate of Field Size Distributions for Selected Sites in the Major Grain Producing Countries.* Greenbelt, MD: Goddard Space Flight Center (X-923-76-93).

Richason, B.F. (editor). 1982. *Remote Sensing: An Input to Geographic Information Systems in the 1980s* (Proceedings, Pecora VII Symposium). Falls Church, VA: American Society of Photogrammetry.

Robinove, C. 1981. "The Logic of Multispectral Classification and Mapping of Land," *Remote Sensing of Environment* 11:231-244.

Robinson, A.H. 1982. *Early Thematic Mapping in the History of Cartography.* Chicago: University of Chicago Press

Schweitzer, G.E., 1982. "Airborne Remote Sensing," *Environmental Science and Technology* 16:338A-346A.

Scott, F.M. et al. 1972. *Oklahoma Land Use and Activity Code.* Oklahoma City: Office of Community Affairs and Planning.

Shelton, R.L. and E.E. Hardy. 1974. "Design Concepts for Land Use and Natural Resource Inventories and Information Systems," pp. 517-537 in *Proceedings, 9th International Symposium on Remote Sensing of Environment. Ann Arbor: University Of Michigan.*

Simonett, D.S. and J.C. Coiner. 1971. "Susceptibility of Environments to Low Resolution Imaging for Land-Use Mapping," pp. 373-394 in *Proceedings, 7th International Symposium on Remote Sensing of Environment.* Ann Arbor: University of Michigan.

Stamp, L.D. 1951. "Land Use Surveys with Special Reference to Britain," pp. 372-393 in T.G. Taylor (editor), *Geography in the Twentieth Century.* London: Methuen.

Stamp, L.D. 1960. *Applied Geography.* Baltimore: Penguin.

Stamp, L.D. 1962. *The Land of Britain: Its Use and Misuse.* London: Longman.

Stamp, L.D. 1966. *A Glossary of Geographical Terms.* London: Longman.

Swain, P.H. and S.M. Davis. 1978. *Remote Sensing: The Quantitative Approach.* New York: McGraw-hill.

Swain, P.H. et al. 1981. "Contextual Classification of Multispectral Image Data," *Pattern Recognition* 13:429-441.

Thomas, I.L. 1980. "Spatial Postprocessing of Spectrally Classified LANDSAT Data," *Photogrammetric Engineering and Remote Sensing* 46:1201-1206.

Thrower, J.W. 1968. *Man's Domain: A Thematic Atlas of the World.* New York: McGraw-Hill.

Todd, W.J. et al. 1980. "Landsat Wildland Mapping Accuracy," *Photogrammetric Engineering and Remote Sensing* 46:509-520.

Tom, C. et al. 1978. *Spatial Land Use Inventory, Modeling, and Projection/Denver Metropolitan Area with Inputs From Existing Maps, Airphotos, and Landsat Imagery.* NASA Technical Memorandum 79710. Greenbelt, MD: Goddard Space Flight Center.

Tregear, T.R. 1958. *A Survey of Land Use in Hong Kong and the New Territories.* Bude, UK: Geographical Publications, Ltd. and Hong Kong: University Press (The World Land Use Survey, Monograph 1).

Tucker, C.J. 1979. "Red and Photographic Infrared Linear Combinations for Monitoring Vegetation," *Remote Sensing of Environment* 8:127-150.

Van Genderen, J.L. and B.F. Lock. 1977. "Testing Land-Use Map Accuracy," *Photogrammetric Engineering* 43:1135-1137.

Van Valkenburg, S. 1949. "A World Inventory," *Economic Geography* 25:237-239.

Van Valkenburg, S. 1950. "The World Land Use Survey," *Economic Geography* 26:1-5.

Vink, A.P.A. 1975. *Land Use in Advancing Agriculture.* New York: Springer-Verlag.

Virginia Citizens Planning Association. 1980. *The Comprehensive Plan.* Richmond: Virginia Department of Housing and Community Development (Community Planning Series No. 3).

Wallis, H. 1981. "Land Use Mapping in the 1980's: The History of Land Use Mapping," *The Cartographic Journal* 18:45-48.

Webster, R. and P.H.T. Beckett. 1968. "Quality and Usefulness of Soil Maps," *Nature* 219:680-682.

Wray, J.R. 1983. "U.S. Geological Survey Uses New Computer Graphics Techniques to Enhance Satellite Maps," *Computer Graphics News* 2, 5:8,9,15,20.

Recommended Readings

Anderson, J.R. 1982. "Land Resources Map Making From Remote Sensing Products," pp. 63-72 Johannsen and Sanders 1982.

Bailey, R.G. et al. 1978. "Nature of Land and Resource Classification — A Review," *Journal of Forestry* 76:650-655.

Bowden, L.W. 1975. "Urban Environments: Inventory and Analysis," pp. 1815-1880 in R.G. Reeves (editor), *Manual of Remote Sensing, Vol II.* Falls Church, VA: American Society of Photogrammetry.

Card, D. 1982. "Using Known Map Category Marginal Frequencies to Improve Estimates of Thematic Map Accuracy," *Photogrammetric Engineering and Remote Sensing* 48:431-439.

Christian, C.S. 1952. "Regional Land Surveys," *Journal of the Australian Institute of Agricultural Sciences* 18:140-146.

Clark, A. 1976. "The World Land Use Survey," *Geographica Helvetica* 31:27-28.

Clawson, M. and C. L. Stewart. 1965. *Land Use Information: A Critical Survey of U.S. Statistics Including the Possibilities for Greater Uniformity.* Baltimore, MD: Johns Hopkins Press.

Congalton, R.G. et al. 1981. *Analysis of Forest Classification Accuracy.* Houston: NASA LBJ Space Center, AGRISTARS RR-U1-04066 JSC-17123.

Congalton, R.G. et al. 1982. *Accuracy of Remotely Sensed Data: Sampling and Analysis Procedures.* Blacksburg, VA: School of Forestry and Wildlife Resources, Virginia Polytechnic Institute.

Darby, H.C. 1970. "Domesday Book — The First Land Utilization Survey," *Geographical Magazine* 12:416-423.

Davis, W.A. and F.G. Peet. 1977. "A Method of Smoothing Digital Thematic Maps," *Remote Sensing of Envirnonment* 6:45-47.

Finn, R.W. 1973. *Domesday Book: A Guide.* London: Phillimore.

Fitzpatrick-Lins, K. 1978. "An Evaluation of Errors in Mapping Land Use Changes for the Central Atlantic Regional Ecological Test Site," *Journal of Research, U.S. Geological Survey* 6:339-346.

Fitzpatrick-Lins, K. 1978a. "Accuracy and Consistency Comparisons of Land Use and Land Cover Maps Made From High-Altitude Photographs and Landsat Multispectral Imagery," *Journal of Research, U.S. Geological Survey* 6:23-40.

Fitpatrick-Lins, K. 1978b. "Accuracy of Selected Land Use and Land Cover Maps in the Greater Atlanta Region, Georgia," *Journal of Research, U.S. Geological Survey* 6:169-173.

Fitzpatrick-Lins, K. 1980. *The Accuracy of Selected Land Use and Land Cover Maps at Scales of 1:250,000 and 1:100,000.* U.S. Geological Survey *Circular* 829.

Fitzpatrick-Lins, K. 1981. "Comparison of Sampling Procedures and Data Analysis for a Land Use and Land Cover Map," *Photogrammetric Engineering and Remote Sensing* 47:343-351.

Hay, A. 1971. "Sampling Designs to Test Land Use Map Accuracy," *Photogrammetric Engineering* 45:529-533.

Kostrowicki, J. (editor). 1965. "Land Utilization in East-Central Europe, Case Studies," *Geographia Polonica 5.*

Loelkes, G.L. 1977. *Specifications for Land Cover and Associated Maps.* U.S. Geological Survey Open File Report 77-555. Reston, VA: USGS.

Loelkes, G.L. and B.A. McCullough. no date. *Ozarks Pilot Land Use Data Base Test and Demonstration Final Report.* Little Rock, AK: Ozarks Regional Commission.

Milazzo, V.A. 1980. *A Review and Evaluation of Alternatives for Updating U.S. Geological Survey Land Use and Land Cover Maps.* U.S. Geological Survey *Circular* 826. Washington, DC: USGPO.

Nunnally, N.R. 1974. "Interpreting Land Use From Remote Sensor Imagery," pp. 167-187 in J.E. Estes and L.W. Senger (editors), *Remote Sensing: Techniques for Environmental Analysis.* Santa Barbara, CA: Hamilton.

Paine, D.P. 1981. *Aerial Photography and Image Interpretation For Resource Management.* New York: Wiley.

Paludan, C.T. 1976. "Land Use Surveys Based On Remote Sensing From High Altitudes," *Geographica Helvetica* 31:17-24.

Peplies, R.W. 1976. "Cultural and Landscape Interpretation," pp. 483-507 in J. Lintz and D.S. Simonett (editors), *Remote Sensing of Environmeeeent.* Reading, MA: Addison-Wesley.

Rhind, D. and R. Hudson. 1980. *Land Use.* New York: Methuen.

Rosenfield, G. 1982. "Sample Design for Estimating Change in Land Use and Land Cover," *Photogrammetric Engineering and Remote Sensing* 48:793-801.

Sauer, C.O. 1921. "The Problem of Land Classification," *Annals,* Association of American Geographers 11:3-16.

Strahler, A.H. 1980. "The Use of Prior Probabilities in Maximum Likelihood Classification of Remotely Sensed Data," *Remote Sensing of Environment* 10:135-163.

Wake, W.H. and G.A. Hull (editors). 1981. *Field Study for Remote Sensing: An Instructor's Manual.* NASA Conference Publication 2155. Washington, DC: NASA.

Wiedel, J.W. and R. Kleckner. 1975. *Using Remote Sensor Data for Land Use Mapping and Inventory — A User's Guide.* Springfield, VA: National Technical Information Serivce (PB-242 813/AS).

Wooten, H.H. 1953. *Major Uses of Land in the United States.* Washington, DC: U.S. Department of Agriculture, *Technical Bulletin* 1802.

Wooten, H.H. and J.R. Anderson. 1957. *Major Uses of Land in the United States — Summary for 1954.* Washington, DC: U.S. Department of Agriculture, *Agriculture Information Bulletin* 168.

Young, H.E. and E.G. Stoeckler. 1956. "Quantitative Evaluation of Photo Interpretation Mapping," *Photogrammetric Engineering* 22:137-143.